Springer Aerospace Technology

The series explores the technology and the science related to the aircraft and spacecraft including concept, design, assembly, control and maintenance. The topics cover aircraft, missiles, space vehicles, aircraft engines and propulsion units. The volumes of the series present the fundamentals, the applications and the advances in all the fields related to aerospace engineering, including:

- structural analysis,
- aerodynamics,
- aeroelasticity,
- aeroacoustics,
- flight mechanics and dynamics
- orbital maneuvers,
- avionics,
- systems design,
- materials technology,
- launch technology,
- payload and satellite technology,
- space industry, medicine and biology.

The series' scope includes monographs, professional books, advanced textbooks, as well as selected contributions from specialized conferences and workshops.

The volumes of the series are single-blind peer-reviewed.

To submit a proposal or request further information, please contact: Mr. Pierpaolo Riva at pierpaolo.riva@springer.com (Europe and Americas) Mr. Mengchu Huang at mengchu.huang@springer.com (China)

The series is indexed in Scopus and Compendex

Marco Evangelos Biancolini · Corrado Groth ·
Ubaldo Cella
Editors

Digital Twins for Aerospace

CAE Modelling of Future Mobility

Springer

Editors
Marco Evangelos Biancolini
Department of Enterprise Engineering
University of Rome Tor Vergata
Rome, Italy

Corrado Groth
Department of Enterprise Engineering
University of Rome Tor Vergata
Rome, Italy

Ubaldo Cella
Department of Enterprise Engineering
University of Rome Tor Vergata
Rome, Italy

ISSN 1869-1730 ISSN 1869-1749 (electronic)
Springer Aerospace Technology
ISBN 978-3-032-05021-2 ISBN 978-3-032-05022-9 (eBook)
https://doi.org/10.1007/978-3-032-05022-9

This Springer imprint is published by the registered company Springer Nature Switzerland AG
The registered company address is: Gewerbestrasse 11, 6330 Cham, Switzerland

Preface

Aerospace engineering is a very young discipline. Despite being born only at the beginning of the last century, it has experienced an exceptional growth compared to many other disciplines. Fields like civil or mechanical engineering have evolved almost steadily over centuries. Aeronautics emerged rapidly in the early-twentieth century and advanced at a remarkable pace. Only few decades passed between the first Wright brothers' flight and the first supersonic jet. Sixty-six years after the flight of the Flyer, a man was able to walk on the surface of the moon. In the meantime aerospace became the main technological driver of all engineering fields. The advent of computers, advanced simulations, and materials science further contributed revolutionizing the field, placing aeronautical engineers at the forefront of technological breakthroughs. Since the second half of the last century engineers have smoothly shifted from manual design and prototyping of aircrafts to computer-aided processes allowing them to test and refine designs virtually before anything is built. Aircraft design has become faster, cheaper and safer year after year. Digital transformation is often cited as the fourth industrial revolution (Fig. 1).

Fig. 1 Visualization of the flow field around an A380

In more recent times, environmental concerns have become increasingly dominant, guiding technological choices and research directions. Industries are pushed to innovate by exploring new, greener solutions (e.g. hybrid or fully electric propulsion systems), and more efficient configurations. Numerical tools are then required to satisfy ever-increasing needs. The field has become more and more multidisciplinary, blending knowledge from AI, software engineering, and materials science. In this scenario Digital Twins (DTs) play a leading role.

Several definitions of the concept of Digital Twin have been proposed. The most concise is: *A Digital Twin is a virtual representation of a connected physical asset.* According to the Members of the AIAA Digital Engineering Integration (DEIC) Committee, a Digital Twin is defined as [1]:

> *A set of* ***virtual information*** *constructs that mimics the* ***structure, context and behavior*** *of an* ***individual/unique physical asset****, or a group of physical assets, is* ***dynamically updated*** *with data from its physical twin* ***throughout its life cycle*** *and* ***informs decisions*** *that* ***realize value.***

This definition broadly applies to a wide range of engineering applications. The DT concept is generally applied with different nuances depending on the discipline involved and its constituent elements. One of the added values of a Digital Twins is the potential to provide a real-time, high-fidelity virtual model of its physical counterparts.

This book presents a selection of academic and industrial works representative of different applications of Digital Twins in aerospace. It also includes research studies focused on aspects related to the accuracy of numerical models. The book begins with an historical review of the aerospace science emphasizing the evolution of design environments in the recent decades. Problems of numerical optimization, both structural and aerodynamic, are discussed in chapters "Structural Optimization of a Wing for Wind Tunnel Testing Through Mesh Morphing" and "A Mixed Intrusive and Non-intrusive Approximation Technique for Efficient Robust Aerodynamic Shape Design". The problem of identifying accurate formulations for the drag components in numerical models is faced in the chapter "Induced Drag and Vorticity in Viscous and Inviscid Flows". A Digital Twin that provides real-time performances of an aircraft is presented in "Enhancing Aircraft Design and Performance Through High-Fidelity Digital Twins: A CFD-Based Approach" while in "An Interactive Design Tool for NACA Intakes Based on High Fidelity CFD Simulations and Reduced Order Models" a Digital Twin is implemented to enhance the design of a NACA intake. The chapter "Accurate Mapping in Fluid-Structure Interaction Analysis Across Models of Varying Fidelity Levels" addresses the problem of accurate loads mapping in FSI analyses. The numerical-experimental activities carried out within the EU research project MOthIF are reported in the chapter "Digital Twin Concept Application: The MOthIF Case". The last paper of the book reviews the role of CAE technologies within the research activities of the Italian National Institute for Astrophysics (INAF).

Rome, Italy

Marco Evangelos Biancolini
Corrado Groth
Ubaldo Cella

Reference

1. Digital Twin: Definition & Value. *An AIAA and AIA Position Paper December 2020*, (https://www.aia-aerospace.org/wp-content/uploads/Digital-Twin-Institute-Position-Paper-December-2020-1.pdf).

Contents

From the Drawing Board to the Digital Twin: The Last 40 Years of Aerospace Engineering

Antonio Maglione and Ubaldo Cella

Abstract Aerospace is a young field of engineering. It is common to date the origin of its industries back to 1903 with the first flight of the Wright brothers' machine over a sandy camp near Kitty Hawk in North Carolina. Nevertheless, it showed a rapid and extraordinary growth that is without precedent in history. There is a generation of people who were able to witness both the first flight of the "heavier than air" and the first man walking on the lunar surface. Such impressive progress was possible thanks to the parallel advancements in technology, materials, and design methodologies. The latter, in particular, benefitted from the equally exceptional evolution of computing capability. This paper traces the principal historical milestones of the aeronautical technology and offers an incomplete review of the evolution of the design methodologies in its industry during the last decades. Some case studies representative of typical design approaches adopted by aircraft industries during this period are also offered.

Keywords History of CAE · Aerospace engineering evolution · Technology progress · Computing advancement

1 Introduction

The dream of flight has always belonged to humanity. Just think of the myth of Icarus in which the young son of Daedalus, overwhelmed by the thrill of flight, lost his mind and his life. However, it was only in the ninth century, during the Islamic Golden Age, that the first attempt of flight was documented in Cordoba (Spain). The poet, inventor and astronomer Abbas Ibn Firnas invented a gliding device which

A. Maglione
Vere Novo Technologies, Naples, Italy

U. Cella (✉)
Department of Enterprise Engineering, University of Rome "Tor Vergata", Rome, Italy
e-mail: ubaldo.cella@designmethods.aero

Design Methods—Aerospace, Messina, Italy

M. E. Biancolini et al. (eds.), *Digital Twins for Aerospace*, Springer Aerospace Technology, https://doi.org/10.1007/978-3-032-05022-9_1

managed to be airborne for a short duration of time [1]. In the 15th centuries people began to think that flying was possible imitating the action of birds. The first who tackled the problem with a scientific approach was Leonardo Da Vinci. In the *Codex on the Flight of Birds* [2], written from the 1505, he discussed the relationship between the centre of gravity and the centre of lifting pressure on a bird's wing, explained the behaviour of birds as they ascend against the wind (foreshadowing the modern concept of a stall), and demonstrated rudimentary understandings of the relationship between cambered wing sections and lift [3]. In the seventeenth century the Italian Jesuit, mathematician and naturalist Francesco Lana de Terzi published a book titled *Prodromo* which contained the description of a "flying ship". He was the first to develop an idea for a "lighter than air" vessel. For the mathematical and scientific rigor with which he faced the study he is considered the founder of aeronautical science [4]. The first rigorous study of the physics of flight is due to Sir George Cayley between the eighteenth and the nineteenth century [5]. His studies, together with the extensive (unlucky) glider experiments of Otto Lilienthal in the late nineteenth century, contributed to laid the groundwork for modern aeronautics. The turning point arrived with Orville and Wilbur Wright. They collected the available literature about aeronautics from the *Smithsonian Institution* and deeply studied their predecessors. They produced many design concepts and validated them against tests but, above all, they tuned and corrected the equations after having concluded that the Lilienthal's data were in error. They developed and instrumented wind tunnels (Fig. 1) to produce measurements that were used to correct the *Smeaton's coefficient* [6] in the lift and drag equations. They tested as many as 200 wing models and collected the aerodynamic data to build a successful flying machine [7]. The Flyer's twelve-second flight, the morning of December 17, 1903, changed the world. The history of aeronautics began and transformed transportation, culture and warfare forever.

Fig. 1 Replica of the Wright brothers' wind tunnel (*Wikipedia*)

The flight of the *Flyer* was the result of the intuition, passionate curiosity and perseverance of two brilliant old hand engineers. But above all it was the success of a method. They studied the theory, deepened both successes and failures of their predecessors and conducted meticulous research to refine solutions and to validate design tools. Their approach influenced the evolution of the new science leading man to walk on the moon surface after just over half a century. Aerospace research has become the main scientific driver of most fields of engineering, which today draw on its technology to innovate.

The progress of aeronautics ran in parallel with the extraordinary progress of computer science and software engineering. The increasing power of "computing machines" further stimulated the evolution of design methodologies. The first computation instruments were purely mechanical machines able to perform elementary arithmetic operations. The first programmable mechanical computer, the *Z1*, dates back to 1936. The first programmable electronic digital computer was completed in 1945. The *ENIAC* (Electronic Numerical Integrator And Computer), this was its acronyms, is often associated to the beginning of the computer age. It was able to process about 500 floating-point operations per second (FLOPS). It weighed 30 tons and was housed in a 140 m^2 room (Fig. 2). Fifty years later, the University of Pennsylvania celebrated its anniversary reimplementing the machine using integrated circuit technology fitting the room-sized computer in the palm of a hand. The supercomputer *LEONARDO*, hosted by *CINECA* (Italy) and classified at its debut (in the end of 2022) in 4° position in the Top500 list, is able to process 240 PetaFLOPS, 5×10^{14} more than *ENIAC*. To have a more practical feeling on the computing power increase it is enough to cite that the Apollo Guidance Computer (AGC), installed onboard of the *NASA* modules used in the lunar missions, was 140 times slower than the pocket *Texas Instruments TI-73* (a calculator very popular in the late 90s and targeted towards school pupils aged 11–14 years old), and had a millionth of the processing power of a modern mobile phone [8].

The growth of computational power induced a vision change in engineers. The role of computers began soon to turn from hardware supporting human calculations to

Fig. 2 *ENIAC* in building 328 at the *Ballistic Research Laboratory* (U. S. Army Photo)

devices used to implement sophisticated design tools. Such shift began in the latter half of the twentieth century with the introduction of the CAE (Computer-Aided Engineering) philosophy. Computers began to be used to develop tools able to model geometries and to simulate complex physical phenomena as structural response and aerodynamics. The new paradigm has revolutionized the aircraft design process. Currently, aerospace engineering relies extensively on CAE technologies in which engineers engage in immersive design experiences, utilizing tools that facilitate and expedite the exploration of optimal designs. These tools enable the recognition and analysis of design performance through extensive simulations, made possible by the Digital Twins of the aircraft itself, in the various critical disciplines and technology segments. Looking ahead, the ambition is to further advance the field by integrating emerging technologies such as machine learning, Artificial Intelligence (AI), virtual and augmented reality into the aircraft design.

This paper provides an incomplete overview of the transformation design environments experienced in aeronautics, reporting a list of milestones that represent the impact the evolution of the design methodologies had on its path. The history of the *Piaggio P.180* will be reported as representative of the Italian technology in the aerospace industries from the 80s. A set of typical aeronautical design problems, together with the numerical approach adopted, will be introduced. An eye on the cutting-edge design technologies will complete the picture.

2 From the Flyer to the Raider

Since the Wright brothers' *Flyer* made its historic flight covering 120 ft in 12 seconds in 1903 at Kitty Hawk, aircraft engineering has relentlessly pushed technological boundaries. The flight fever spread throughout the world, exciting and approaching people to what was considered a new sport. Aeronautics became a spectacular exhibition of modernity and technology that, nevertheless, maintained, at its begin, a handcraft character (Fig. 3). Several eccentric inventor-pilots, eager to emulate the feats of the two American brothers, were stimulated to try their hand at building unlikely prototypes that barely managed to take off. This pioneering moment marked the dawn of powered flight and set the stage for a century of innovation and progress in aviation. It became immediately clear, in fact, that the aircraft was a revolutionary means with significant industrial, commercial and military implications. Over the decades, advancements in materials, aerodynamics, propulsion systems, and avionics have transformed aircrafts from fragile, wood-and-fabric contraptions into sophisticated, high-performance machines. In just three decades, the relatively short and uncertain flight of the *Flyer* was surpassed by remarkable achievements in aviation technology.

The First World War, during which the important strategic and tactical role of the airplane emerged, gave a great impulse to its development in a more structured way. The several achievements of the post-war period bear witness to this extraordinary evolution. It was a period pervaded by daring challenges, one of the most important of which was the feat of Charles Lindbergh (Fig. 4) with his first non-stop solo flight

Fig. 3 The flight of the Wright *Flyer II* (Library of Congress)

across the Atlantic ocean. In the early morning of May 20, 1927, Lindbergh took off from Long Island, near New York, flew for 5800 km and landed in Le Bourget, near Paris, the evening of May 22, after 33 h and 32 min. Although the ocean had already been flown over (but not with a solo flight) in 1919 by the British aviators John Alcock and Arthur Whitten Brown, the Lindbergh's flight with its *Ryan NYP Spirit of St. Louis* marked a significant step in opening to the intercontinental routes.

Fig. 4 Charles A. Lindbergh standing by his plane (Library of Congress)

Fig. 5 *Macchi-Castoldi M.C.72* (Aeronautica Militare)

On the technological side, the evolution of aircrafts was further stimulated by aviation competitions as the Schneider Cup, a race that became particularly popular during the twenties and early thirties. The prestigious deriving from winning the Schneider Trophy pushed governments to finance the development of innovative prototypes able to break speed records. One of the most notable milestones was the *Macchi M.C.72* (Fig. 5) developed in this contest, a seaplane that, in 1934, exceeded an astonishing speed of 709.21 km/h [9]. This piston-powered marvel was a testament to the rapid advancements in aerodynamics, engine design, and materials science that characterized the early twentieth century. The *Macchi M.C.72*'s achievement not only set a new benchmark for speed but also demonstrated the potential for innovation and progress in the field of aviation.

In the same period of time, on July 14, 1937, the *Tupolev ANT-25* (Fig. 6), a Soviet long-range experimental aircraft, made a historic non-stop flight from Moscow to San Jacinto, California. This remarkable journey covered a distance of approximately 11,500 km [10]. This flight was a testament to the aircraft's engineering and the crew's skill and posed a significant milestone in the history of aviation. The primary objective of this ambitious flight was to foster plans for regular mail services between the United States and the USSR. By establishing a direct air route over the North Pole, the two nations aimed to enhance communication and strengthen diplomatic ties. The *Tupolev ANT-25*'s achievement showcased the potential for international cooperation in aviation and highlighted the strategic importance of polar routes for global connectivity.

The crew, consisting of pilot Valery Chkalov, co-pilot Georgy Baidukov, and navigator Alexander Belyakov, faced numerous challenges during the flight, including extreme weather conditions and navigation difficulties. Their successful completion of the mission was a proof to their bravery, determination, and expertise. The flight of the *Tupolev ANT-25* remains a symbol of innovation and progress in the field of aviation.

Fig. 6 *Tupolev ANT-25* (San Diego Air and Space Museum Archive)

The *DC-3* (Fig. 7), designed by Arthur Emmons Raymond, first flew in 1935, only thirty-two years after Wright Brothers took off the *Flyer*. The *DC-3* is still one of the most remarkable airplanes in the history of aviation thanks to an all-metal, low-wing monoplane design, aluminium fuselage with a wide cabin allowing for comfortable seating for up to 32 passengers. Thanks to its advanced features, within two years *DC-3* captured the market for land-based airliners [10]. Built to last, to land anywhere and never to break, *Douglas Company* produced more than 16,000 *DC-3*s and *C-47*s, its military sibling, before and during World War II. Remarkably the almost 90-year-old *DC-3* continues to fly today with an estimated 172 aircraft still in operation across all variants [11].

With an outstanding speed exceeding Mach 3.2, the *Lockheed SR-71 Blackbird* (Fig. 8) is still the fastest air-breathing manned aircraft ever built. Designed by

Fig. 7 The *Douglas DC-3* (RAeS/NAL)

Fig. 8 *Lockheed Martin SR-71 Blackbird* (*NASA*)

Clarence "Kelly" Johnson and his famous Skunk Works team, known for delivering "impossible" technologies on incredibly short, strategically critical deadlines. The Blackbird took the skies in 1964, just six decades after the Wright' *Flyer* and operated until its retirement in 1998 after more than thirty years of successful service. The design team had to invent the vast majority of the technologies used in this remarkable aircraft. It was constructed primarily from titanium to withstand the extreme temperatures generated at high speeds and was powered by two *Pratt & Whitney J58* engines which could operate in afterburner mode for extended periods. The *Blackbird* could operate at altitudes of 85.000 ft (26,000 m) allowing it to evade threats by flying higher than most missiles and aircraft [12].

Slightly slower than the *Lockheed SR-71 Blackbird*, the *Concorde* (Fig. 9) remains the fastest commercial aircraft ever built. It could cruise at speeds of up to Mach 2.04 (2179 km/h) and at altitudes of 60,000 ft (18.3 km), reducing the flight time between London and New York to about three hours. The *Concorde* first flew in December 1967, just six-and-a-half decades after Kitty Hawk *Flyer* enterprise, and was the result of a joint venture between the *British Aircraft Corporation* (*BAC*) and *Sud-Aviation*, a French state-owned aircraft manufacturer, laying the foundation for *Airbus Industrie*s [10].

The first *Boeing 747-100* (Fig. 10) completed is first test flight in 1969, revolutionizing air travel as the world's first twin-aisle airplane, enabling more people to fly farther, faster and more affordably than ever before. Marked by its distinctively recognizable hump, this iconic airplane is a symbol of great engineering and innovation. The *B.747* is often noted as an outstanding work of architecture. Since

Fig. 9 *Concorde* (*Wikipedia*)

production, began in 1967, more than 100 customers have purchased 1574 aircraft, logging more than 118 million flight hours and nearly 23 million flight cycles.

Following the continuous development of technologies fuelled by the aircraft industry during WWII, aviation history took a significant turn in the 1950s. On October 4, 1957, approximately half a century after the first motorized flight, the Soviet Union successfully launched *Sputnik I*, the world's first artificial satellite. Measuring about 58 cm in diameter and weighing 83.6 kg. The *Sputnik I* (Fig. 11) marked the start of the space age, orbiting the Earth on its elliptical path in about 98 min.

Building on the successes of earlier space missions, Yuri Gagarin made history on April 12, 1961, by becoming the first human to journey into outer space and orbit the Earth, after only three and a half years after the launch of the *Sputnik 1*. Gagarin's spacecraft, *Vostok 1* (Fig. 12), was launched by the Soviet Union from the Baikonur Cosmodrome in Kazakhstan. The mission lasted approximately 108 min, during which Gagarin completed one orbit around the Earth at an altitude of about 327 km.

Fig. 10 The last *Boeing 747* ever produced (*Wikipedia*)

Fig. 11 The first artificial satellite *Sputnik 1* (*ESA*)

Gagarin's achievement was a significant milestone in the Space Race between the United States and the Soviet Union, demonstrating the Soviet Union's advanced capabilities in space exploration. The success of *Vostok 1* not only marked a major technological and scientific accomplishment but also had a profound impact on global geopolitics and the future of human spaceflight.

The American public was shocked by the success of the Soviet Union. The USA Administration reacted pushing on the *Vanguard* program, the first attempt to launch an American satellite, but it was afflicted by a frustrating disaster. In the wake of this public failure, and in the awareness of Soviet superiority, the US Congress approved a law that turned the *National Advisory Committee on Aeronautics* (*NACA*) into the

Fig. 12 Gagarin's *Vostok 3KA* capsule at the *RKK Energiya* museum (*Wikipedia*)

National Aeronautics and Space Administration (*NASA*). The Army's space-related activities were transferred to *NASA*. The proud reaction of the United States was made official in 1962 by President Kennedy with his famous speech at the Rice University's Stadium in which he announced the *Apollo* program aimed to land an American on the Moon. The rest of the story is known and describes one of the most wonderful and glorious pages in the use of human ingenuity.

The decades after the crewed Moon landing were marked by several success in missions, aimed to the exploration of the solar system, and by the put in orbit, in April 1971, of the first space station: the Soviet *Salyut 1*. However, a protagonist of this period was certainly the *Space Shuttle*, the first partially reusable low Earth orbital spacecraft system. Its program, whose official name was Space Transportation System (STS), was taken in 1969 from a plan already active during the *Apollo* program. Six orbiters were built for flight. The first fully functional one was *Columbia* which was launched April 12, 1981 (Fig. 13). The *Space Shuttle* operated from 1981 to 2011, flew 135 missions and carried 355 astronauts from 16 countries.

Remarkably, all those outstanding aircraft and spacecraft designs, along with their contemporary counterparts, were conceived and developed in a largely "analogue" engineering world without, or with extremely limited, help of some rudimental computing device. A symbol of this pioneering era of aircraft engineering is

Fig. 13 Launch of the *NASA Space Shuttle Columbia* (*NASA*)

the *slide rule* (Fig. 14), a hand-operated mechanical device engineers used for evaluating mathematical operations such as multiplication, division, exponents, roots, logarithms, and trigonometry. During this time, hundreds of new designs each year expanded the aircraft flight envelope in speed, altitude and load capability, as well as computing spacecraft trajectories around the Earth, with their designers aided by just a slide rule!

Modern symbols of the era of digital design are the successes of Space X, in the field of space exploration, and the Airbus 380, in representing the aircraft evolution. Nevertheless, to end up this very synthetic and incomplete picture of the aircraft development timeline, from the Wright brothers' *Flyer* to today's state-of-the-art technology, it is worth mentioning the *Northrop Grumman's B-21 Raider* (Fig. 15). This Sixth-Generation fighter aircraft is set to become the core of the US Air Force's strategic bomber fleet and is currently in the critical flight-testing phase. The *Raider* features advanced stealth technology, next-generation networking capabilities and an open systems architecture, integrating data, sensors, and weapons to deliver unprecedented capability and flexibility.

Dubbed "a digital bomber", the *B-21* utilizes agile software development, advanced manufacturing techniques and digital engineering tools to reduce production risks and enhance modern sustainment practices. It employs Cloud Technology for ground systems data and includes a digital twin to support operations and sustainment for a more maintainable and sustainable aircraft with lower infrastructure costs. Networking together manned and unmanned systems, the *B-21 Raider* will incorporate new levels of integration teaming with a swarm of "loyal wingman" unmanned aircraft to move forward into high-risk areas without exposing "human" flying crew.

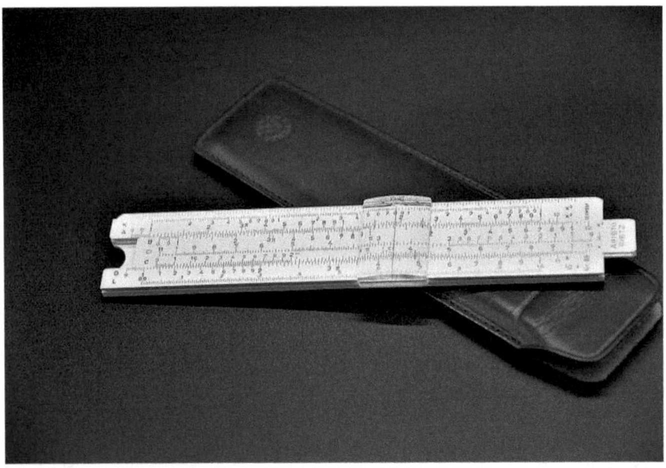

Fig. 14 "Log-log" slide rule

Fig. 15 *Northrop Grumman B-21* flight test aircraft (USAF)

2.1 Availability of Computing Power

After the pioneering era of early computing machines, where Alan Turing's work laid the foundation for modern computing and had a profound impact on the outcome of World War II, computer science slowly and reluctantly entered in aircrafts design. The first digital computers specifically used for this purpose was the *Ferranti Mark I* (Fig. 16), the world's first commercially available general-purpose electronic computer, designed at the University of Manchester by Freddie Williams and Tom Kilburn [13].

Based on the *Manchester Mark I* and built by *Ferranti Ltd.*, the huge machine was housed in two spacious bays, each 5 m long, 2.4 m high, and 1 m wide, with a control desk at one end. Inside were 4000 electronic valves (or vacuum tubes), 2500 capacitors, 15,000 resistors, and nearly 10 km of wiring. It consumed 27 kW of power and operated at a base clock frequency of 100 kHz. The *Ferranti Mark I*'s unique memory system could store just over a kilobyte of data in cathode-ray tubes for high-speed access. A rotating magnetic drum offered 82 kB of permanent storage.

Fig. 16 Alan Turing (standing) on a *Ferranti Mark I* console

Fig. 17 The last flight of the last *Avro Vulcan* in military service (2015)

The *Ferranti Mark I*, introduced in early 1951, was a groundbreaking development in the world of computing and represented a significant leap forward in technology. *Avro*, a prominent British aircraft manufacturer, recognized the potential of this powerful machine and employed it during the design of advanced aircrafts, such as the *Avro Vulcan* (Fig. 17). The *Ferranti Mark I*'s computational capabilities allowed Avro's engineers to perform complex calculations and simulations, enabling them to optimize the design and performance of the aircraft. This innovative use of computing power accelerated the development process and contributed to the creation of more efficient and sophisticated aircrafts.

Computing power was popularized by the *IBM System/360* (Fig. 18). Announced in 1964, the *System/360*, often referred as "*The 5-billion-dollar gamble that changed the trajectory of IBM*", was a system of mainframe that ushered in a new era of compatibility in which computers were no longer thought of as collections of individual components, but rather as platforms. *IBM* introduced Solid Logic Technology (SLT), which packed more transistors onto a circuit card, allowing for more powerful yet smaller computers. Designed to cover both commercial and scientific applications, the *System/360* was versatile and widely adopted. For the first time, software written for one machine could run on any other machine in the line. Its success allowed customers to purchase smaller systems with the confidence that they could expand them as needed without reprogramming application software or replacing peripheral devices.

Due to the large computational costs associated with the solutions of the math at the basis of its theory, aerodynamic design is one of the disciplines that has particularly benefited from computing power. As the computing machines became more powerful, they played a crucial role in the development of Computational Fluid Dynamics (CFD) allowing engineers to simulate airflow over wings and entire aircraft. *NASA*'s *Ames Research Center* has been a pioneer in advancing computational methods for

Fig. 18 *IBM System/360* operator's console, and peripherals at *Volkswagen* (*Wikipedia*)

aircraft design. The use of supercomputers, like the *Cray-2* (Fig. 19), introduced in 1985 as the fastest supercomputer in the world at the time showing a peak performance of 1.9 gigaflops, revolutionized the field by enabling more detailed simulations and analyses. This immense computational power, for that time, allowed researchers to perform complex simulations that were previously impossible, helping engineers understanding how air flows around aircraft surfaces.

The symbiotic relationship between computers and wind tunnels allowed designers to process wind tunnel test results more efficiently and accurately, and to predict some of those results before a model was even built. Wind tunnel data,

Fig. 19 *Cray 2* supercomputer (*NASA*)

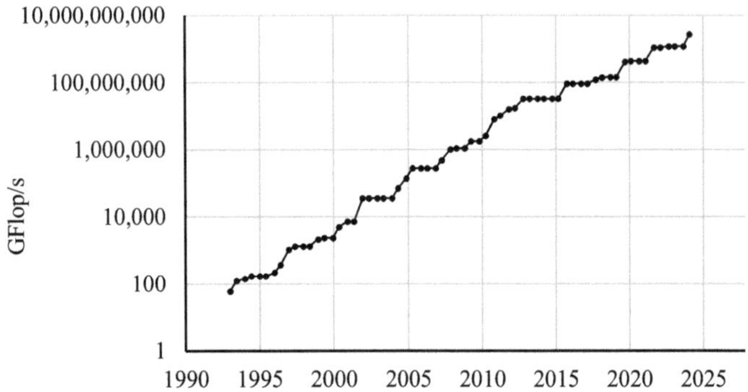

Fig. 20 Growth of supercomputers performances from 1993 [14]

in turn, enabled designers to develop better programs for their computers to predict airflow.

In the late 80s, supercomputers became massively parallel machines often based on thousands of processors. Their CPUs were initially explicitly developed for supercomputing but at the end of the 90s, the transition to the use of generic processors, providing high performance at reduced costs due to mass production, was completed. In recent years, supercomputers have become increasingly parallel machines equipped with an ever-increasing number of processors. At the date of November 2024, the most powerful supercomputer in the world is *El Capitan—HPE Cray EX255a*, equipped with more than 11 million cores and able to reach a theoretical peak performance of 2746.38 PFlop/s. The evolution of the supercomputer performance during the last 32 years is plotted, in logarithmic scale, in Fig. 20.

The most recent architectures also use processors that collaborate with graphics cards. Graphics Processing Units (GPU), in fact, have proven to have very high-performance for some types of mathematical operations typical of scientific computation. Modern supercomputers use batteries of high-performance graphics cards. Hardware based on GPUs acquired strategic importances in the recent years due to rising interest in Artificial Intelligence. Such components are, in fact, ideal for accelerating heavy AI workloads.

3 Review of CAE Technology Evolution

The term Computer-Aided Engineering is attributed to Dr. Jason R. Lemon, founder of *SDRC*, in the late 70s. It describes a generic field that include the use of technologies as solid modelling, structural analysis, computational fluid dynamics, computational electromagnetics, multibody dynamics and numerical optimization, to support engineers in analyses and design processes. The main protagonist of this field has historically been the CAD (Computer-Aided Design) although it is often referred to a discipline separated from CAE. With the evolution of numerical analysis methods, and with the introduction of the parametric geometric modelling paradigm, CAE technologies began to be integrated into multidisciplinary environments extending the spectrum of technologies. Typical examples are Fluid-Structure Interaction (FSI) analyses, Multidisciplinary Optimization (MDO) and, not least, Digital Twins.

The first two-dimensional rudimentary CAD systems were developed during 60s. *Sketchpad*, written by Ivan Sutherland in 1963, was the first graphical system that allowed users to interact with the program through a screen (Fig. 21) [15]. It is considered the progenitor of CAD programs. In the 70s researches began slowly to include the third dimension. Several developing programs began to appear. In this contest it was soon evident also the need to identify a standard of communication between different technologies. The initiative was taken by the United States Air Force within the *Integrated Computer Aided Manufacturing* (ICAM) project. The standard, named IGES (Initial Graphics Exchange Specification), was first published in March, 1980 by the U.S. National Bureau of Standards [16]. Few years later (1984) a second-generation of technology was started and submitted to ISO (International Organization for Standardization) in 1988. The international community adopted it as the basis for ISO 10303 [17], informally known with the acronyms STEP (which stands for "Standard for the Exchange of Product model data").

The path toward three-dimensional solid modelling began to consolidate during 80s. A milestone in this direction was the invention of NURBS (Non-Uniform Rational Basis-Splines) by Versprille [18] which laid the mathematic basis of 3D curves and surfaces modelling. In 1987, *Parametric Technology Corporation* released the commercial *Pro/ENGINEER* in which features were described by

Fig. 21 *TX-2* operating area—*Sketchpad* in use [15]

double-precision solid geometry and NURBS surfaces. *Pro/E* also introduced the revolutionary parametric feature-based solid modelling approach. The inclusion of 3D rendering capability marked the start of the modern era of CAD systems.

In aircraft industry the investment in CAD systems began in 1965 when *Lockheed* started an internal project aimed to write a software for *IBM* mainframes. In 1972 the corporation, beset by economic difficulties, was induced to exploit the business opportunity in selling its design software which acquired the name *CADAM* (Computer-graphics Augmented Design and Manufacturing). *Dassault* was one of the first aircraft industries to purchase the software which, at that time, was limited to two-dimensional capabilities. The potentialities of the tool and the significant benefits deriving from its improvement, as well as in applying interactive graphics, were soon recognized. Stimulated by the needs of the recent introduced Numerical Control (NC) machines, the company started the development of a three-dimensional design tool which was later renamed in *CATIA* (Computer-Aided Three-Dimensional Interactive Application). In 1981 a separate subsidiary, called *Dassault Systèmes*, was established to commercialize the software with *IBM*. In 1986 *Boeing* selected *CATIA* as design software. October 1990 the company launched the twin-engine 777 family program using *CATIA* V3 for drawing. *Boeing* adopted a conservative approach supporting the digital design with the traditional generation of a set of physical mock-ups. The test on the first mock-up, generated for the nose of the aircraft, was so successful that the build of all other planned mock-ups was cancelled. The *Boeing* 777 became the world's first 100% digitally designed aircraft [19] marking a definitive breakthrough in the design environment of aircraft industries. It was a successful strategic choice that demonstrated the importance of innovation in design methodologies. Such innovations, nevertheless, can create overconfidence leading to neglect the opportune cautions. An example of how critical such decisions might be refers to the A380 program. In March 1998, *Dassault Systèmes* previewed *CATIA* Version 5 which marked a technology break with the past. *Airbus* introduced the use of the new version of the software but not in all sites of the company. The German and Spanish design teams maintained the use of *CATIA* V4 while the French and British offices had upgraded to *CATIA* V5. This choice caused that the German teams could not add their design changes for the electrical wiring of A380 back into the common 3D digital mock-up being produced in Toulouse involving a mismatching of cables length during assembling parts produced in different sites. Hundreds of small changes were necessary which snowballed the program into a year's delay and causing a loss probably close to €5 billion, or $6.6 billion, in profit over the following four years for *EADS* [20]. Figure 22 shows the evolution of the Graphical User Interface (GUI) of the *CATIA* CAD system from its first version to the V5 release.

Almost simultaneously to the first CAD systems, software focused on simulation began to be developed. The first computer-aided simulation technologies were mainly oriented on Finite Element Analysis (FEA), despite the first documentation on the use of computer to face numerically a fluid dynamic problem appeared in the early 50s [21]. What we consider the earliest form of the finite element method was developed by Argyris, at the *Imperial College* London, and M. J. Turner, at *Boeing* Company,

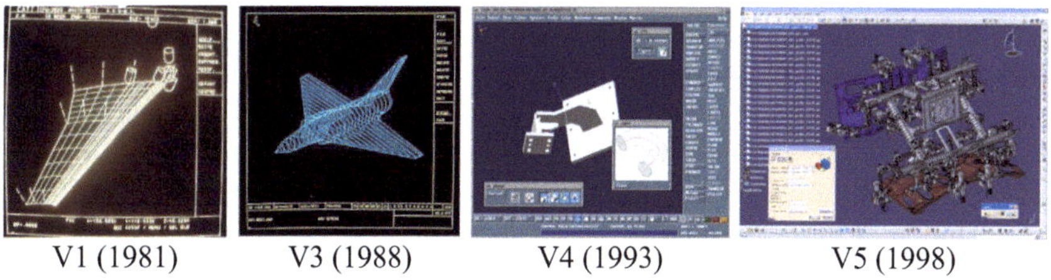

Fig. 22 UI screenshots of historical releases of *CATIA*

in 1954 [22]. At that time the method was called *Matrix Stiffness Method*. The code run on a computer with 4k of 16-bit memory and could solve a maximum of approximately 40 equations. Turner applied the new method of structural analysis at *Boeing* contributing to the development of successful programs as the *Boeing* 727. The term *Finite Element Method* (FEM) was coined by Dr. Ray Clough. It appeared in 1956 in the first published paper on this topic [23] while the first book devoted to FEM appears in 1967 [24]. In the early 1960s, *NASA* awarded the *MacNeal-Schwendler Corporation* (MSC) with a contract to develop a general-purpose FEA code. The solver, called *NASTRAN* (*NASA* Structural Analysis) was later distributed to the public as open source while *MSC* continued independently the development of its software which in 1969 became the first commercial FEA solver. One year later John Swanson releases the first version of his *ANSYS* (ANalysis SYStems) software. Hibbitt, Karlsson, and Sorensen released the first version of *ABAQUS* in 1978. The development and improvement of graphics processing, during 80s, opened FEA methods to *pre* and *post* processing capabilities contributing to a significant simplification of software usage. Analysts could graphically setup problems and visualize coloured solution speeding up the analyses and deepening the evaluation of the solutions. FEA methods began to be extensively incorporated into product design processes. Those years were called *The Golden Age* of the Finite Element Method. Today numerical structural analysis is a mandatory task in aerospace industries and it is involved in the certification process of aircrafts [25].

A strong impulse to solve fluid dynamic problems arrived after the Whitcomb's experiments on supercritical wings which led *NASA* to assign the task of developing a computational method for designing supercritical airfoils to the *Courant Institute* at New York University. In 1974 Antony Jameson's developed the *FLO22* code for 3-D swept wing design [26]. This moment represents a kind of turning point between a pre-CFD era, mainly centred on linear-potential theory, and the ERA-1 in which the implementation of methods was based on the solutions of nonlinear potential equations (established by the theoretical work of Murman and Cole [27] in 1970). ERA-2, approximately identified in the last two decades of the twentieth century, was characterized by the diffusion of Euler and RANS (Reynolds-averaged Navier-Stokes) solvers. The basis of the theory refers to the work of Jameson et al. [28]. The end of the last century saw a rapid rise of the commercial software industry. The first commercial general-purpose CFD code was *PHOENICS* (acronym for Parabolic,

Hyperbolic Or Elliptic Numerical Integration Code Series) launched to the market in 1981 [29]. The first CFD solver having a graphical user interface and workflow rather than a command-line input was *FLUENT* which became the dominant market leader by the late 90s. An extensive review of this period, from a European perspective, and of the application of CFD for aircraft design, is reported in [30]. ERA-3 saw the evolution of unsteady flow solvers with hybrid RANS/Large-Eddy-Simulation (LES) techniques marking the modern age of CFD. A milestone on this topic is the theoretical work of Spalart et al. [31]. The above division of the evolution of CFD was suggested by Rizzi and Luckring [32] which also related it to the evolution of supercomputing (Fig. 23).

Early CAE software and hardware were extremely expensive making them accessible only to large corporations and research institutions. The poor computational resources available at the begin limited their use to rough calculations of simple domains. The systems were very complex and easily affected to programming errors. In the early days, users had to create input decks consisting of hundreds or even thousands of punched cards, discovering errors only after the solver crashed. It was even more ridiculous when the cards eventually scattered everywhere on the floor after an accidental drop of the boxes, very much like a humoristic comedy scene! The development of interactive interfaces significantly contributed to simplify CAE usage and to speed up setups, at the same time the continuing decline in the costs of computer hardware made the economic scenario more favourable and contributed to expand the number of players in the market of software. Nevertheless, the real breakthrough

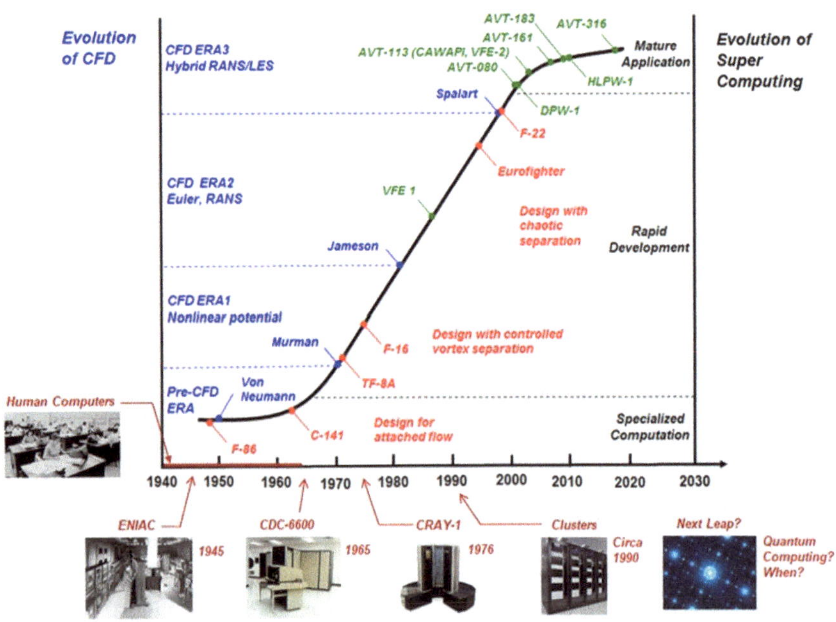

Fig. 23 Evolution of CFD in relation to the evolution of computing power [32]

could occur during 80s with the introduction of personal computers. This was the moment R&D office environments converged to the path of definitive retirement of drawing board. Desks layout began to assume their final configuration around PCs and workstations and engineers adapted their competences toward a new design workflow.

As technology progressed, CAD and CAE software started to converge, leading to tighter integration between design and analysis tools. This integration enhanced collaboration between designers and analysts, streamlining the iterative design process. Nevertheless, it also suffered by an historical cultural isolation of disciplines that slowed down the process. The complexity and generally obscure nature of CAE tools restricted the simulation capability to experts who, on the other hand were often poorly opened to disciplines other than own. A typical example is the interaction between aerodynamicists and design engineer. The requirements of CAD and CAE based design process are regularly conflicting involving a costly iteration often hampered by a lack in competences overlapping. Things remained this way for decades until parametric CADs consolidated the replacement of direct modelling tools. In a parametric model, features are associated with attributes such as length, width, depth, position, orientation etc., and references to other features [33]. The model entities can then be linked each other. The full model can be regenerated from the history tree without the need of repeating building workflow. This approach is philosophically and methodologically significantly different from the traditional modelling in which each feature is an isolated "history free" feature. The introduction of this new technology caused, in some cases, a kind of generational change in experienced CAD users that were not able to upgrade to a new design methodology but also opened new potentialities in the development of CAE technologies. A vision change took hold involving the formation of a new generation of engineers opened to a multidisciplinary approach. At the begin of 2000s the costs and the availability of computer power suggested to investigate the possibility of integrating Multi-Disciplinary Optimization approach in the industrial design process.

Numerical optimization was not a new discipline. In the field of structures, the first example was established by Schmit in 1960 which coupled a structural analysis with a numerical optimization criterion [34]. Few years later he presented the first known multidisciplinary design optimization application [35] involving aerodynamics and structures. Aerodynamic shape optimization began in 1974 with Pironneau [36]. He used optimal control techniques to minimize the drag of a body by varying its shape via "control" variables. In 1988 Jameson optimized aircraft wings extending the adjoint method with more sophisticated CFD models [37]. The following years saw the improvement of decision-making algorithms based on several approaches [38]. The market began to offer commercial software able to more friendly support the coupling of numerical models to develop optimization environment also offering a variety of optimization algorithms. At the begin of twenty-first century technologies and hardware were mature enough to stimulate the diffusion of multidisciplinary design optimization at industrial level.

The modern perspectives in design methodologies include the integration of machine learning and artificial intelligence (AI). AI-powered algorithms can analyse vast datasets, identify patterns, and generate design solutions that might not be apparent through traditional methods alone unlocking innovative configurations. In combination with efficient Reduced Order Models (ROM) and compression algorithms able to manage large data sets is possible to create digital twin that enables real-time monitoring, analysis, and optimization of performance throughout an aircraft's operational life. By utilizing data from sensors and onboard systems, engineers can make informed decisions to enhance safety, reliability, and efficiency.

4 Italian Aircraft Masterpiece from the 80s

During the oil crisis of the 1970s, exacerbated by the closure of the Suez Canal occurred after the start of the Six-Day War, the aircraft industry *Rinaldo Piaggio S.p.A* (today *Piaggio Aerospace*), one of the world's oldest aircraft manufacturers, embarked on an ambitious project to develop an aircraft that would marry the speed of jet planes with the fuel efficiency of propeller aircrafts. Drawing inspiration from their heritage of previous aircraft models, the company's designers focused on achieving optimal aerodynamic performance [39]. Key features identified included a mid-wing arrangement to minimize aerodynamic interference with the fuselage, a laminar wing to reduce cruise drag, and the use of pusher propellers to avoid turbulent airflow over the wing. This innovative approach aimed to create an aircraft that was not only fast but also highly efficient in fuel consumption. However, these characteristics resulted in a relatively short fuselage, leading to significant centre of gravity limitations due to the aft-mounted turboprop nacelle. This is where the ingenuity of the Italian designers truly shone. They extended the fuselage forward of the wing front spar, meticulously selecting its length, height, and volume to comfortably accommodate 6 to 9 passengers. Balancing this arrangement using a traditional aerodynamic configuration proved to be impossible. Therefore, a small wing was ingeniously added to the fuselage nose to shift the aerodynamic centre of effort, ensuring a stable configuration. This patented "3-lifting-surface" solution revolutionized the design. The entire airplane configuration development process is well represented by the wind-tunnel models (Fig. 24), which are still preserved and were tested in the company's facility (Fig. 25). This facility was designed in the 1930s by Professor Giuseppe Gabrielli, a Sicilian engineer that studied under Theodore von Kármán and was one of the most eminent designers of his era [40].

Fig. 24 Some of the historical wind tunnel models of *P.180* (*Piaggio Aerospace*)

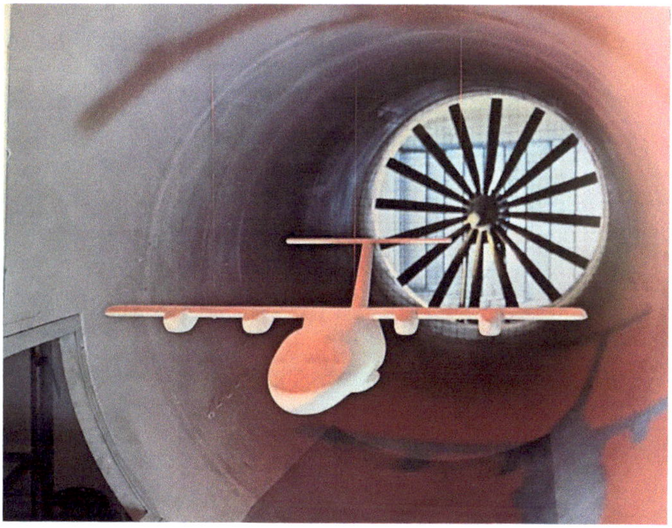

Fig. 25 *Piaggio* historical wind tunnel testing facility in Finale Ligure (*Piaggio Aerospace*)

This innovative approach marked a significant milestone in aeronautical engineering and showcased the brilliance and forward-thinking of the minds involved, setting a new standard for future designs. The models themselves prove the meticulous work and groundbreaking vision that went into creating a more efficient aircraft. This clever design is at the heart of the unique appearance of the *P.180*. In fact, its commercial name, "*Avanti*," is derived from the Italian word meaning "forward," highlighting its forward-thinking design.

All this remarkable engineering work was carried out in the late seventies and early eighties, predominantly using analogue tools. Despite this, the design process for the *P.180 Avanti* continues to evolve even today. This ongoing development is driven by several necessities, including the obsolescence of equipment, evolving airworthiness requirements, and performance enhancements made possible by technological advancements. For instance, the *P.180* configuration has been upgraded with several new features (Fig. 26). These include a low-noise setup using low-RPM scimitar blade propellers with specific exhaust ducts, and winglets that enhance the climb rate and range. All these design processes have been executed using state-of-the-art digital tools, representing a smooth engineering transition from an aircraft originally conceived with analogue methods to a modern, competitive product thriving in the digital age.

Building on the commercial success of the *P.180 Avanti*'s redesign in the early 2000s, the company seized the opportunity to expand its product line by introducing a new family of business jets, dubbed the *P.1XX* (Fig. 27). This strategic initiative aimed to penetrate a more lucrative segment of the market, enhancing the company's presence and competitiveness.

Fig. 26 *Piaggio P.180 Avanti EVO (Piaggio Aerospace)*

Fig. 27 *Piaggio P.1XX (Piaggio Aerospace)*

Right from the start, this design effort was conducted in a fully digital environment, leveraging state-of-the-art tools within an integrated research and development process. This digital approach allowed for greater precision, efficiency, and innovation throughout the design phase. A prime example of this methodology was the dedicated project to identify the optimal aerodynamic configuration for a laminar-flow wing. This configuration was meticulously optimized not only for superior aerodynamic performance but also for maximizing usable fuel storage volumes and streamlining manufacturing processes. The integration of advanced digital tools enabled the design team to explore various design iterations rapidly, ensuring that the final product met both performance and market demands. The laminar-flow wing, in particular, was a significant achievement, as it contributed to reduced drag and improved fuel efficiency, making the *P.1XX* a highly competitive offering in the business jet market. Furthermore, the digital design environment facilitated seamless collaboration among different engineering teams, allowing for the integration of cutting-edge technologies and innovative solutions. This holistic approach ensured that every aspect of the *P.1XX*'s design was scrutinized and refined to meet the highest standards of performance, safety, and efficiency. Unfortunately, the program was not developed further, but the engineering of the *P.1XX* family of business jets marked a significant milestone for the company, showcasing their ability to innovate and adapt to market demands while leveraging the latest digital tools and technologies.

Based on the capacity achieved with *P.1XX*, the engineering team developed an Unmanned Air System based from *P.180 Avanti* platform in the early years of the last decade, aimed at Intelligence, Surveillance, and Reconnaissance (ISR) missions (Fig. 28). Leveraging their extensive knowledge of the aerodynamics of the unique 3-lifting-surface configuration, acquired over thousands of flight tests, the design team quickly developed a 6 Degree of Freedom mathematical model. This model was pivotal in the development process, especially for creating flight control laws encoded in safety-critical software. This solid engineering foundation instilled great confidence in the control laws, which performed flawlessly. This success was due in large part to extensive testing conducted in proprietary flight simulators. These simulators allowed the crew to meticulously test parameter evolution and rehearse various flight phases, ensuring a smooth transition to actual flight testing.

Recently, the authors have been engaged in the conceptual design phase of a groundbreaking seaplane. This innovative design incorporates a hull enhanced with hydrofoils, aimed at significantly improving landing performance in varying sea states. The addition of these foils helps to stabilize the seaplane during take-off and landing, allowing it to operate effectively even in rough water conditions. To address

Fig. 28 *Piaggio P.1HH Hammerhead (Piaggio Aerospace)*

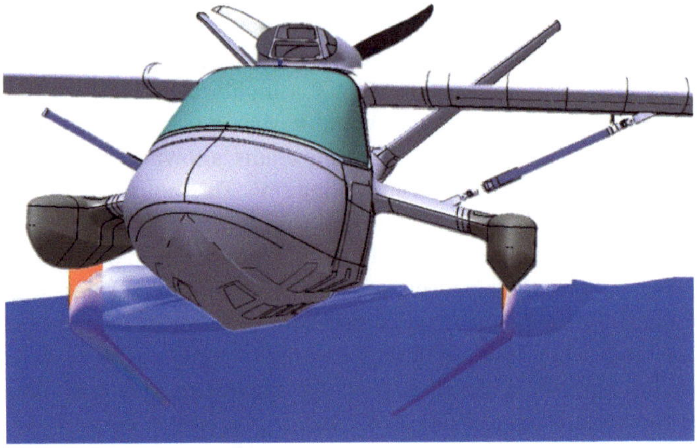

Fig. 29 Numerical analysis of foils to be installed on the *Novotech Seagull*

the complexities introduced by this novel configuration, a pioneering modelling approach is required. This approach must accurately simulate the six degrees of freedom (6DoF) that the seaplane experiences during its transition from floating to full flight mode. This involves detailed computational models to predict how the seaplane will behave as it moves from the water to the air, ensuring seamless integration and optimal performance of the hydrofoils.

The modelling process leverages advanced computational fluid dynamics and other state-of-the-art digital tools to provide a comprehensive understanding of the aerodynamic and hydrodynamic interactions (Fig. 29) [41]. By simulating these interactions with high fidelity tools, the seaplane's performance could be optimized ensuring it meets rigorous safety and efficiency standards.

The integration of hydrofoils represents a significant leap in seaplane technology, promising enhanced operational capabilities and greater versatility. This new approach not only addresses traditional challenges associated with seaplane design but also paves the way for future advancements in amphibious aircraft engineering.

5 Examples of Case Studies from the Last Decades

During 80s CAE technologies were able to support experimental tests with analyses of single configurations of mechanical systems. FEA tools were used to evaluate the stress state of a structure and CFD offered the possibility to visualize the flow field around relatively simple geometries providing preliminary estimation of the performances to be validated against experiments. A pivotal milestone in the use of CAE tools was the advent of Multi-Disciplinary Optimization during 90s. The new approach allowed for the simultaneous consideration of numerous design factors, such as aerodynamics, weight distribution, and structural integrity. This holistic approach led to more balanced designs and improved performance characteristics. The application of CAE advanced technologies, nevertheless, remained for long time limited to research environments. The integration in the industrial process of aircrafts design occurred when cost-effective computational resources and reliable numerical tools reached a sufficient maturity. In this scenario, the combination of research toward the integration within an industrial program offers the major opportunities to gain potential advantage to the marked. An example of realization of such perspective is the Italian research project *VITAS* (*Vettore Innovativo per il Trasporto Aereo Sostenibile*) started in 2001 by *Piaggio Aerospace* in the view of launching a new innovative product. The project, conducted in collaboration with *CIRA* (Italian Aerospace Research Centre), was aimed to develop numerical tools for aerodynamic laminar design and to manufacture a transonic Natural Laminar Flow (NLF) wing to be tested at operative Reynolds numbers. The design of the transonic NLF wing run in parallel to the progress of the *P.1XX* program, briefly introduced in the previous section. It covered four years in which the research activity was strongly integrated with the multidisciplinary design process of the aircraft. In [42] the description of the wing design, highlighting the complexities associated to the fulfilment of the requirements deriving from the several design disciplines involved (structural requirements, systems allocation, manufacturing, fuel storage capacity, etc.), is described. The success of the *VITAS* project was certified by three weeks of experimental tests in the pressurized *DNW–HST* transonic wind tunnel in Amsterdam [43] (Fig. 30).

As already introduced, numerical optimization is a discipline that significantly contributed to improve design capabilities. Antony Jameson pioneered this field for aerodynamic design in the late 80s. Its application in the industrial design process become a common practice in the early 2000s. In Fig. 31 an example applied to the design of winglet is reported. This work, that dates back to 2008, is based on coupling

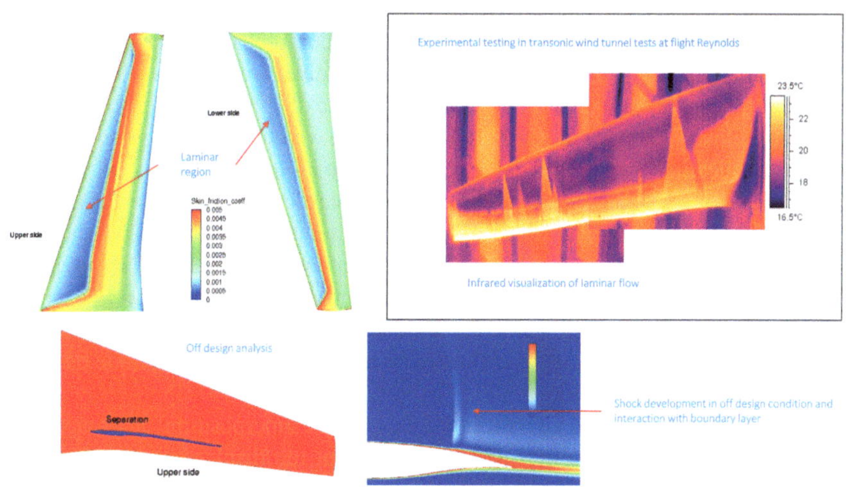

Fig. 30 Numerical solutions and measurements on the *VITAS* transonic NLF wing

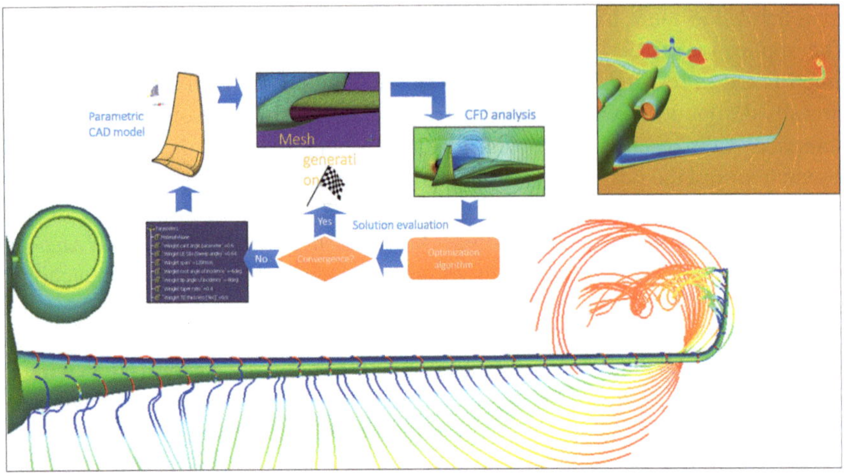

Fig. 31 Sketch of a CAD-based numerical optimization workflow of winglets

a parametric CAD model with a CFD automatic analysis [44]. The workflow was made efficient thanks to the adoption of a remeshing procedure based on a multiblock structured hexahedral topology able to adapt to the parametric CAD model.

One of the critical aspects of numerical procedures that involves analyses of non-fixed geometries, is the strategy to be used to adapt the numerical domain to new shapes. In the previous example this problem was approached by a remeshing procedure based on a multiblock structured mesh. The complexity of such setups significant rises with the complexity of the geometry, limiting its practical application to relatively simple topologies. For this reason, in most industrial problems, unstructured meshes are adopted. However, a remeshing approach is, in these cases, often impracticable due to the high computational costs related to the generation of unstructured fluid dynamic domains. Several alternatives are possible. Among them, mesh morphing has proven to be one of the most flexible and robust approaches. A multidisciplinary adoption of CAE technologies in which mesh morphing found its natural application is the Fluid-Structure Interaction analysis. In [45] the so called 2-way FSI procedure is applied to the static aeroelastic analysis of the *P.1XX* wind tunnel model (Fig. 32). The method consists in iterating between FEM and CFD solutions up to the convergence to the final shape of the structure under aerodynamic loads. Key points of the method are the implementation of the two solvers coupling and the strategy to adapt both numerical domains to the deformed shape of the wing. The latter task was, in this case, performed by a mesh morphing technique based on Radial Basis Functions (RBF).

The FSI problem can be also faced by the "modal superposition" approach. The idea at its base is the simplifying assumption that the displacement under loads can be replicated by a combination of a limited number of structural natural modal shapes. This method can be very efficiently implemented adopting mesh morphing offering the possibility to generate an intrinsically aeroelastic CFD model. In [46]

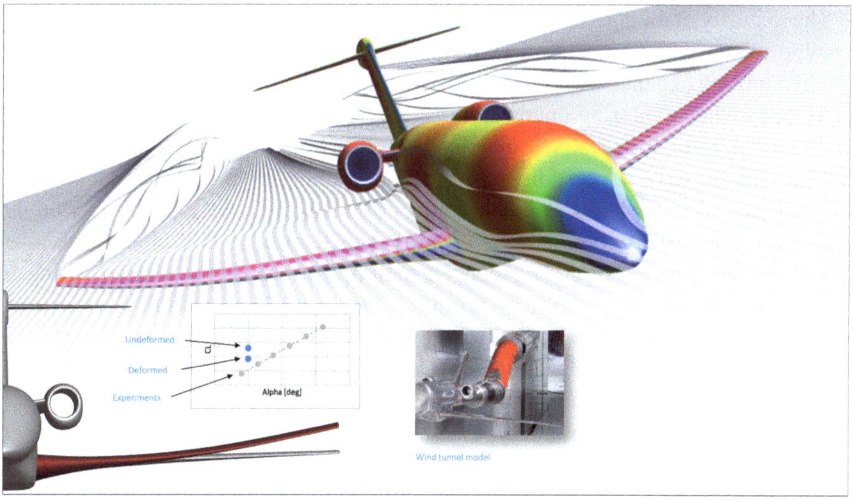

Fig. 32 Static fluid-structure interaction (FSI) analysis of a wind tunnel model

the implementation of a modal FSI analysis is described and compared to the 2-way FSI solution.

The aerodynamic optimization example reported in Fig. 31 did not account for the flexibility of the structure. This aspect is particulary felt in designing winglets. A more modern approach to design combines the response of the structure runtime within the aerodynamic analysis. An example of a numerical aeroelastic analysis, implemented setting-up the modal superposition approach, is following reported. The case study is referred to the optimization of the winglets of the *P180 Avanti EVO*. The structural model included the wing box, a limited section of the fuselage and a wireframe with concentrated masses modelling the engine (Fig. 33). The analysis procedure is an evolution of tools developed in the framework of the EU funded RBF4AERO project [47].

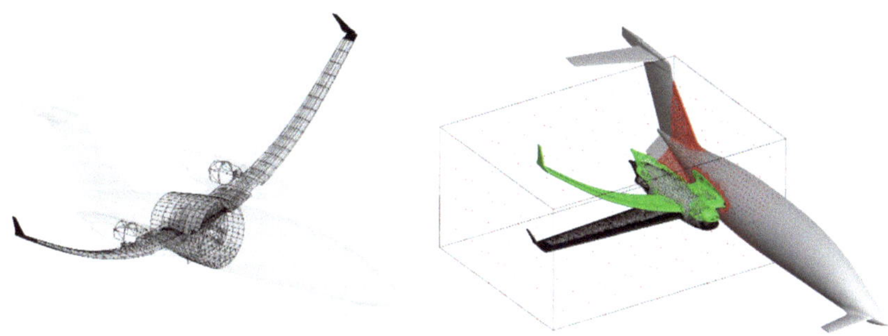

Fig. 33 Structural model of the *P.180* and RBF setup of the mesh morphing problem

The maturity of modern CAE tool and the confidence in the accuracy of the several possible numerical approach, allow to investigate the performance of preliminary designs limiting expensive experimental campaigns. This aspect aquires particularly importance when exploring unconventional configuration. In [48] numerical setups based on "high fidelity" RANS solvers and a "low-fidelity" Vortex-Lattice Method (VLM), suitable to be adopted in preliminary aircraft design process, are descripted. The methods were applied to study the effect of installing a tractor propeller on the wing tip (Fig. 34). This configuration recently gained apeal in the contentex of novel hybrid electrical propulsion for greener aviation. Tip propellers, when rotating in the direction counteracting wing tip vortices, acts in shifting the trailing vortex core outboard enlarging the aerodynamic wing span and increading the efficiency. The study allowed to improve the confidences on the accuracy of several numerical tools and confirmed the validity of the configuration in gaining a drag reduction.

CAE tools offer the possibilty to model complex flow fields and to design components in operative configurations often difficult to replicate experimentally. An example of such case is reported in [49]. The study was aimed to optimize the horizontal empennage of a rotorcraft in operative conditions. The tail-mounted horizontal stabilizer surface is immersed in the wake generated by the main rotor. The flow field around the empennage is then pertubated and depends on the flyght conditions (Fig. 35). A sophisticated CAD driven optimization procedure, using a decision making criterion basen on genetic algorithms, allowed identifying a 3D configuration able to generate a significant improvement in the stability derivative of the empennage.

One of the recent advances in desing methodologies involves the creation of digital twins adopting numerical tecniques to reduce the number of problems variables. The

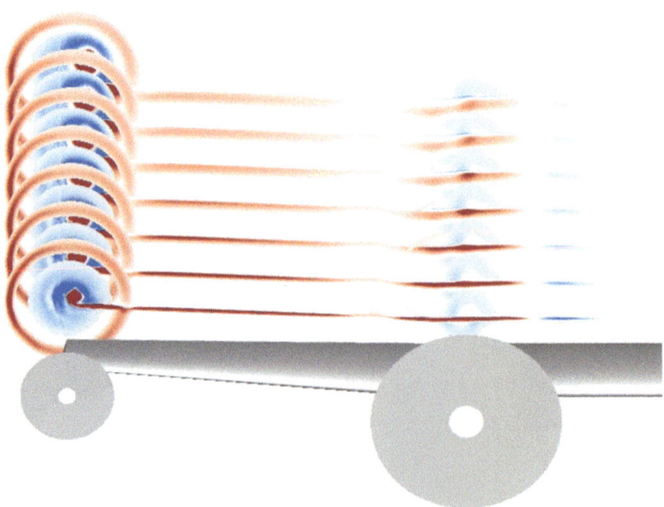

Fig. 34 Effect of tip and traditional propellers on the wing wake

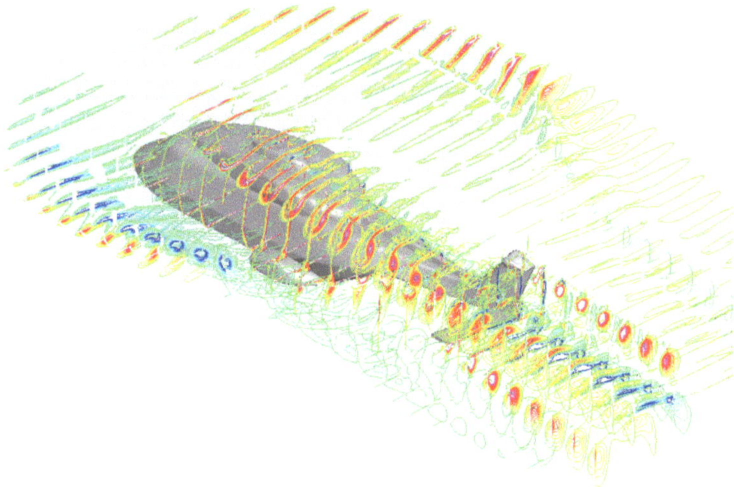

Fig. 35 Flow field around a rotorcraft in operative forward flight

objective is to generate a ROM able to provide a fast solution of a complex flow field in function of a wide variable space. This technique allows to evaluate the effects of changing the problem variables in real time, simplifying the identification of the optimal compromising design. In [50] the feasibility of using reduced models for the aeroelastic optimization of a scoop air intake for aeronautical applications is evaluated. The FSI solutions are generated by modal superposition. A decomposition tecnique was used to reduce the number of variables, identifying the modes whose linear superposition allow the reconstruction of the field quantities with a controlled error. Machine Learning algorithms were used to correlate the input parameters to the weights of the principal modes. Figure 36 reports the comparison between the fluid dynamic solution (pressure on the scoop surfaces and duct) computed by CFD and the one estimated by ROM. It is worth mentioning that the errors generated adopting this approach is related to the number of adopted modes and can be constrained.

The reported design examples are partially representative of the evolution in design capabilities during the last decades, period in which commercial aviation industry achieved more than 20 dB of noise and 70% of CO_2 emissions reduction [51]. The perspettive is clearly to progress in this path towar greener, faster,

Fig. 36 Comparison between CFD and ROM pressure field

more efficient, and lighter aircraft. Current challenges include loads alleviation and dynamic FSI analyses of extremely flexible wings with very high aspect ratio. Challenges related to the power of computational resources are the capability to model a complete aircraft in take off condition including all details as landing gears, slats and flap tracks.

6 Conclusions

The short history of aeronautics represents one of the most admirable examples of technological development in human history. In 1903, the short flight of the Wright brothers' machine marked the dawn of human-controlled flight and laid the foundation for the rapid and profound evolution of aviation technology (Fig. 37).

The following years were time of experimentation and refinement in aircraft design toward the development of practical airplanes, which were mostly used for military reconnaissance, civilian airshows and mail delivery. The period between the first and the second world-war marked a golden age of aviation, with a surge in technological advancements and the advent of faster, longer-range aircraft. The monoplane design became increasingly popular due to its aerodynamic efficiency. Commercial aircrafts began to incorporate metal construction, retractable landing gear, and more advanced cockpit instruments for better performance and safety. World War II altered the trajectory of aircraft design demanding for faster and higher-performing

Fig. 37 Wright Brothers' *Flyer* first flight (Library of Congress)

aircraft. This period saw the rapid development of jet-powered aircraft. After the war, commercial aviation also experienced a transformation with the advent of jetliners.

By the late twentieth and early twenty-first centuries, aviation entered in a new phase, characterized by the need for greater efficiency, safety, and sustainability. The increasing complexity of aircraft design, combined with the demands of environmental consciousness, noise impact over the near-by communities and the rising fuel costs, drove the industry to seek innovative solutions through advanced technologies, including materials science, flight management systems, and, most notably, Computer-Aided Engineering tools. The integration of CAE technology in aircraft design has been a transformative journey that has significantly reshaped the aerospace industry. Over the decades, the application of CAE tools has enabled engineers to model, simulate, and analyse aircraft systems with unprecedented precision, reducing the reliance on physical prototypes and streamlining the design as well as the manufacturing process. Moreover, CAE played a crucial role in enhancing the understanding of complex physical phenomena, such as turbulence, shockwave formation, and combustion processes, which were once beyond the reach of traditional design methods.

The early years of CAE were marked by limited computational power, which constrained the complexity of simulations and the accuracy of results. However, as computational resources grew and software tools evolved, CAE became integral to various stages of aircraft development, from structural analysis and aerodynamics to materials selection and fatigue prediction. The transition from traditional hand calculations to digital simulations enhanced both the speed and quality of design iterations, allowing for the creation of more efficient, safer, and lighter aircraft. Over the past four decades, with the increase of computing power, CAE technology has profoundly transformed aircraft design, enhancing efficiency, safety, and performance. CAE systems became extremely integrated with manufacturing processes, streamlining the certification process and reducing time-to-market for new aircraft. Further, the concept of digital twins living in the operational scenario enables real-time monitoring and predictive maintenance.

Today, CAE technology is an indispensable part of the aerospace industry. Looking ahead, the future of CAE in aircraft design will likely be driven by the need for even more sustainable and efficient aircraft. As environmental considerations become increasingly critical, CAE tools will play a central role in the development of greener technologies, such as hybrid-electric propulsion systems and advanced materials with lower environmental impact. Additionally, with the advent of additive manufacturing and more advanced computational capabilities, the role of CAE in designing customized, lightweight, and high-performance components will continue to grow. In the use of CFD for aircraft design, the future perspective will consist in a larger migration from RANS to unsteady Navier-Stokes simulation, (from RANS-LES to full LES) and/or Lattice Boltzmann method. Design strategies will make large use of Artificial Intelligence and Machine Learning to develop Reduced Order Model able to enable real-time decision-making during the design process. Virtual Prototyping and Digital Twins will become more prevalent for in-service aircrafts, enabling continuous performance tracking and monitoring, allowing a performance and safety-oriented model for predictive maintenance. Real-time simulations will allow to test different scenarios and make design adjustments dynamically during testing or operational use.

The history of CAE technology in aircraft design is a testament to the power of innovation and its ability to drive progress in the aerospace industry. The continuous refinement of simulation techniques is expanding the boundaries of what is possible in aircraft design. Several challenges are still on progress. The accurate simulation of separated flows in high lift conditions, for instance, are still open problems for which no sufficiently reliable numerical tools are available. As we look ahead, CAE evolution promises new frontiers in sustainability, automation and design that will once again reshape the future of flight.

References

1. Jamsari, E.A.: Mohd Aliff Mohd Nawi, Adibah Sulaiman, Roziah Sidik, Zanizam Zaidi, and M. Z. A. H. Ashari, Ibn Firnas and his contribution to the aviation technology of the world. Adv. Nat. Appl. Sci. **7**(1) (2013)
2. Zanon, E.: Book of the Codex on Flight, Biblioteca di S.M. il re, Leonardo3, Turin, Italy (2009)
3. Jakab, P.: Leonardo da Vinci and flight, on faculty.fairfield.edu, Fairfield University. Web site visited Feb 2025. https://web.archive.org/web/20210424104423/.faculty.fairfield.edu/jmac/sj/scientists/lana.htm (2013)
4. MacDonnell, J.: Jesuit Geometers: A Study of Fifty-Six Prominent Jesuit Geometers During the First Two Centuries of Jesuit History. Saint Louis Institute of Jesuit Sources, p. 21 (1989)
5. Ackroyd, J.A.D.: Sir George Cayley: the invention of the aeroplane near Scarborough at the time of Trafalgar. J. Aeronaut. Hist. (2011/6) (2011)
6. Anderson, J.D., Jr.: Fundamentals of Aerodynamics, 6th edn. McGraw-Hill Education, New York (2017)
7. Padfield, G.D., Lawrence, B.: The birth of flight control: an engineering analysis of the Wright brothers' 1902 glider. Aeronaut. J. (697) (2003)
8. Kendall, G.: The first moon landing was achieved with less computing power than a cell phone or a calculator. Web site visited Nov 2023. https://psmag.com/social-justice/ground-control-to-major-tim-cook (2019)
9. Bendoni, M.: L'epopea del reparto alta velocità. Rongoni & Gozzoli (1971)
10. Crouch, T.D.: Wings: a History of Aviation from Kites to the Space Age. WW Norton (2004)
11. Royal Aeronautical Society: Web site visited Feb 2025. https://www.aerosociety.com/news/the-immortal-dc-3/
12. Lockheed Martin: Web site visited Feb 2025. https://www.lockheedmartin.com/en-us/news/features/history/blackbird.html
13. Gandy, A.: The Early Computer Industry: Limitations of Scale and Scope. Palgrave Macmillan, London (2012)
14. TOP500: https://www.top500.org
15. Sutherland, I.E.: Sketchpad: A Man-Machine Graphical Communication System. Lincoln Laboratory, Massachusetts Institute of Technology, Technical Report No. 296 (1963)
16. Nagel, R.N., Braithwaite, W.W., Kennicott, P.R.: Initial Graphics Exchange Specification IGES, Version 1.0. National Bureau of Standards, NBSIR 80-1978, Washington DC (1980)
17. Myer, K.: Handbook of Materials Selection. Wiley (2002)
18. Versprille, K.J.: Computer-aided design applications of the rational B-spline approximation form. Ph.D. thesis, Syracuse University (1975)
19. Norris, G., Wagner, M.: Boeing 777. Motorbooks International, St. Paul, Minnesota (1996)

20. Clark, N.: The airbus saga: crossed wires and a multibillion-euro delay. The International Herald Tribune, Dec 11 (2006)
21. Kawaguti, M.: Numerical solution of the Navier-stokes equations for the flow around a circular cylinder at Reynolds number 40. J. Phys. Soc. Jpn. **8**, 747–757 (1953)
22. Liu, W.K., Li, S., Park, H.S.: Eighty years of the finite element method: birth, evolution, and future. Arch. Comput. Methods Eng. **29**, 4431–4453 (2022)
23. Turner, M.J., Clough, R.W., Martin, H.C., Topp, L.J.: Stiffness and deflection analysis of complex structures. J. Aeronaut. Sci. **23**(9), 805–823 (1956)
24. Cheung, Y.K., Zienkiewicz, O.C.: The Finite Element Method in Structural and Continuum Mechanics, 1st edn. McGraw Hill (1967)
25. Selvarathinam, A.S., Rousseau, C., Engelstad, S.P., Flansburg, L.: Role of FEA, closed-form, and empirical models in certifying aircraft composite structures. In: 57th AIAA/ASCE/AHS/ASC Structures, Structural Dynamics, and Materials Conference, AIAA 2016-1232 (2016)
26. Jameson, A., Caughey, D.A., Newman, P.A., Davis, R.M.: A Brief Description of the Jameson-Caughey NYU Transonic Swept-Wing Computer Program FLO 22, NASA TM X-73996 (1976)
27. Murman, E.N., Cole, J.D.: Calculation of plane steady transonic flows. AIAA J. **9**(1), 114–121 (1971)
28. Jameson, A., Schmidt, W., Turkel, E.: Numerical solutions to the Euler equations by finite volume methods using Runge-Kutta time-stepping schemes. AIAA Paper 81–1259 (1981)
29. Spalding, D.B.: A general-purpose computer program for multi-dimensional one- and two-phase flow. Math. Comput. Simul. **23**(3), 267–276 (1981)
30. Vos, J.B., Rizzi, A., Darracq, D., Hirschel, E.H.: Navier–Stokes solvers in European aircraft design. Prog. Aerosp. Sci. **38**(8), 601–697 (2002)
31. Spalart, P.R., Jou, W.-H., Strelets, M., Allmaras, S.R.: Comments on the feasibility of LES for wings, and on a hybrid RANS/LES approach. In: Advances in DNS/LES, 1st AFOSR International Conference on DNS/LES, Grey den Press, Columbus, OH (1997)
32. Rizzi, A., Luckring, J.M.: Historical development and use of CFD for separated flow simulations relevant to military aircraft. Aerosp. Sci. Technol. **117** (2021)
33. Mantyla, M., Nau, D., Shah, J.: Challenges in feature-based manufacturing research. Commun. ACM **39**(2), 77–85 (1996)
34. Schmit, L.A.: Structural design by systematic synthesis. In: Proceedings of the 2nd National Conference on Electronic Computation. American Society of Civil Engineers, New York (1960)
35. Schmit, L.A., Thornton, W.A.: Synthesis of an airfoil at supersonic Mach number, CR 144. National Aeronautics and Space Administration (1965)
36. Pironneau, O.: On optimum design in fluid mechanics. J. Fluid Mech. **64**(01) (1974)
37. Jameson, A.: Aerodynamic design via control theory. J. Sci. Comput. **3**(3), 233–260 (1988)
38. Martins, J.R.R.A., Ning, A.: Engineering Design Optimization. Cambridge University Press (2021)
39. Gavazzi, P.: Volare avanti. Storia degli aerei Piaggio. Proedi Editore (2000)
40. Gabrielli G.: Una vita per l'aviazione. Bompiani Editore (1982)
41. Maglione, A., Cella, U., Biancolini, M.E., Lecce, L.: Innovative approach for use of hydrofoils on ultralight seaplane. Prog. Mar. Sci. Technol. **5**, 87–96 (2020)
42. Cella, U., Quagliarella, D., Donelli, R., Imperatore, B.: Design and test of the UW–5006 transonic natural laminar flow wing. AIAA J. Aircr. **47**(3), 783–795 (2010)
43. Fauci, R., Nicolì, A., Imperatore, B., Cella, U.: Wind tunnel tests of a transonic natural laminar flow wing. In: 25th AIAA Aerodynamic Measurement Technology and Ground Testing Conference, 5–8 June 2006, San Francisco, California, USA (2006)
44. Cella, U., Romano D.G.: Assessment of optimization algorithms for winglet design. EnginSoft Newsletter, Year 7, n°1 (2010)
45. Cella, U., Biancolini, M.E.: Aeroelastic analysis of aircraft wind–tunnel model coupling structural and fluid dynamic codes. AIAA J. Aircr. **49**(2), 407–414 (2012)
46. Biancolini, M.E., Cella, U., Groth, C., Genta, M.: Static aeroelastic analysis of an aircraft wind-tunnel model by means of modal RBF mesh updating. ASCE'S J. Aerosp. Eng. **29**(6) (2016)

47. Bernaschi, M., et al.: The RBF4AERO benchmark technology platform. In: 7th ECCOMAS Congress 2016, 5–10 June 2016, Crete Island, Greece (2016)
48. Minervino, M., Andreutti, G., Russo, L., Tognaccini, R.: Drag reduction by wingtip-mounted propellers in distributed propulsion configurations. Fluids **7**(7), 212 (2022)
49. Minervino, M., Vitagliano, P.L., Quagliarella, D.: Helicopter stabilizer optimization considering rotor downwash in forward-flight. Aircr. Eng. Aerosp. Technol. **88**(6), 846–865 (2016)
50. Lopez, A., Cella, U., Groth, C., Biancolini, M.E.: Design and optimization of aeronautical components and digital twins development. IOP Conf. Ser. Mater. Sci. Eng. **1306** (2023)
51. Chaput, E.: Exascale needs & challenges for aeronautics industry. PRACEdays15, Dublin, May 26 (2015)

Structural Optimization of a Wing for Wind Tunnel Testing Through Mesh Morphing

Emiliano Costa, Stefano Porziani, Marco Evangelos Biancolini, Corrado Groth, Andrea Chiappa, and Ubaldo Cella

Abstract This paper presents the structural optimization of the wing tunnel model designed for experimental aeroelastic measurements within the EU RIBES research project. Pressure data, deriving from CFD simulations, were transferred to the CSM model following a one-way FSI coupling approach thus enabling the structural optimization of the wing. A parametric study was performed by varying critical thicknesses and adjusting the position of the ribs within the wing. Shape modifications were implemented using RBF-based mesh morphing resorting to the RBF4AERO platform. The results of the optimization are described and analyzed.

Keywords FSI · CFD · Load mapping · RBF · Mesh morphing

1 Introduction

Structural optimization plays a crucial role in the aeronautical field significantly impacting design, performance, and sustainability of aircraft [1]. In essence, structural optimization involves refining the geometry, materials, and configuration of aircraft structures to achieve the best possible performance, often balancing competing factors [2]. Structural optimization allows engineers to enhance the aerodynamic performance, improve load and mass distribution, and increase resistance to fatigue and damage, ultimately leading to longer service life and reduced maintenance burden. A fundamental tool in this process is the parametric study, which provides a comprehensive approach to explore and quantify the effect of various design parameters [3]. By leveraging parametric studies, engineers can efficiently navigate complex design spaces, identify key factors affecting performance, and, ultimately, steer the design towards optimal solutions [4]. This method not only

E. Costa (✉)
ENGYS, Trieste, Italy
e-mail: e.costa@engys.com

S. Porziani · M. E. Biancolini · C. Groth · A. Chiappa · U. Cella
University of Rome Tor Vergata, Rome, Italy

M. E. Biancolini et al. (eds.), *Digital Twins for Aerospace*, Springer Aerospace Technology, https://doi.org/10.1007/978-3-032-05022-9_2

enhances the effectiveness of optimization, but also provides valuable insights that contribute to innovative and robust aeronautical designs.

One-way Fluid-Structure Interaction (FSI) is a simplified approach where the influence of aerodynamic forces on the structure is considered, but the structural response does not feedback to affect the aerodynamic flow [5]. In this method the loads, usually derived from Computational Fluid Dynamics (CFD) simulations in the form of pressure and shear stress distributions, are applied to the structure to assess its performance under expected operational conditions. These aerodynamic loads are transferred to the Computational Structural Mechanics (CSM) model, with accuracy depending on the chosen algorithm [6]. The structural analysis then evaluates the response of the structure in terms of deformations, stresses, and buckling. The deformations calculated are not used to update the CFD aerodynamic model. One-way FSI is particularly useful and beneficial in early-stage design and optimization where rapid iterations are needed to refine and screen design concepts. As a matter of fact, it simplifies the more complex and computationally intensive two-way FSI approach by decoupling the fluid and structural domains thus making one-way approach faster and less resource-demanding [7, 8]. Despite its simplifications, one-way FSI provides valuable insights into the structural performance of aircraft components, allowing engineers to optimize for weight reduction, material selection, and structural integrity while ensuring that the design can handle aerodynamic loads effectively. In case of its application to the aeroelastic analysis of swept back wing it is assumed to offer a conservative approach due to the decrease of loads deriving from the incidence reduction of the outer sections when the wing is bent under load. In absence of aerodynamics unsteadiness and torsional responses of the structure, the one-way approach is fully justified for the static FSI analysis of straight wings.

In the current study, we propose a structural optimization methodology based on the exploration of parametric design spaces resulting from the one-way FSI analysis, specifically applied to the wing model used in the wind tunnel test campaign within the RIBES (Radial Basis Functions at Fluid Interface Boundaries to Envelope Flow Results for Advanced Structural Analysis) European project [9]. The project achieved several key milestones [10, 11], including the development of a load mapping procedure [12] and a structural optimization framework, both validated through the experimental wind tunnel campaign [13, 14]. The aim of the tests were, in fact, to validate FSI methodologies by using a representative wing box structure designed to replicate realistic aircraft wing topologies. The wing was dimensioned to maintain internal stresses within the elastic range fulfilling the safety requirements of the facility while allowing as large as possible deformations, thus enabling a more pronounced observation of the FSI effects. Such conflicting requirements made the design process particularly challenging. The optimization approach supported the exploration of an optimal compromising configuration.

The optimization framework was based on the finite element method integrated in a multi-objective optimization approach by response surfaces. Shape modifications were implemented at the grid level by changes imposed as displacements within the structural domain. A careful selection of parameters and constraints ensured smooth grid modifications minimizing the risk of mesh distortion.

The parametric study of the RIBES wing leverages the tools and methodologies developed within the RBF4AERO platform [15], a European project that ran from September 2013 to August 2016 [16]. The RBF4AERO project focused on advancing aircraft numerical design and optimization through the use of a shape optimization tool based on Radial Basis Functions (RBF) interpolation [17]. This innovative tool enabled the creation of parametric CFD models, allowing for high-performance meshless morphing techniques that facilitate rapid and an accurate design exploration of shape variations. By integrating these techniques into the optimization process, the RBF4AERO platform has significantly contributed to more efficient and robust structural design methodologies within the aeronautical industry. Both the RIBES and RBF4AERO projects aimed to provide designers with analytical tools devised to offer the flexibility necessary to meet often conflicting requirements throughout the process of conceptualization and analysis of modern aircraft [18].

2 Radial Basis Functions Interpolation

RBF interpolation [19] is a powerful mathematical technique for approximating and interpolating multivariate functions, especially useful when dealing with scattered data in multiple dimensions [20]. The growing use of RBF over time, often in contexts far different from their original purpose of simple interpolation, demonstrates their capability to adapt to a wide range of problems. The main idea at the base of RBF interpolation is to construct an interpolating function $s(\mathbf{x})$ that passes through a set of given data points $\{(\mathbf{x_i}, f_i)\}$, where $\mathbf{x_i} \in \mathbb{R}^n$ are the input data points, and f_i are the corresponding function (scalar) values at the assigned points. The interpolating function is defined as a weighted sum of radial basis functions, each centered at a data point:

$$s(\boldsymbol{x}) = \sum_{i=1}^{N} w_i \varphi(\|\boldsymbol{x} - \boldsymbol{x}_i\|) \tag{1}$$

Here, w_i are the weights to be determined, $\varphi(\|\boldsymbol{x} - \boldsymbol{x}_i\|)$ is the radial basis function, and $\|\boldsymbol{x} - \boldsymbol{x}_i\|$ denotes the Euclidean distance between the point $\boldsymbol{x}$ and the center $\boldsymbol{x}_i$.

Common types of radial basis functions are listed in Table 1, where $r = \|\boldsymbol{x} - \boldsymbol{x}_i\|$ and is a shape parameter that controls the spread of the basis function [21].

To find the weights w_i, we set up a system of linear equations by enforcing the interpolation condition:

$$s(\mathbf{x}) = \sum_{i=1}^{N} w_i \varphi(\|\mathbf{x} - \mathbf{x_i}\|) = f_i \tag{2}$$

Table 1 Most common radial basis functions

RBF	$\varphi(r)$
Spline type	r^n, n odd
Thin plate spline	$r^n \log r$, n even
Multiquadratic	$\sqrt{1+\epsilon^2 r^2}$
Inverse multiquadratic	$\frac{1}{\sqrt{1+\epsilon^2 r^2}}$
Inverse quadratic	$\frac{1}{1+\epsilon^2 r^2}$
Gaussian	$e^{-\epsilon^2 r^2}$

for $j = 1, 2, \ldots, N$.

This system can be written in matrix form as:

$$\mathbf{\Phi w} = \mathbf{f} \tag{3}$$

where $\mathbf{\Phi}$ is the interpolation matrix with elements $\varphi_{ij} = \varphi(\|\mathbf{x}_i - \mathbf{x_j}\|)$, $\mathbf{w}$ is the vector of weights, and $\mathbf{f}$ is the vector of function values. To increase the interpolator accuracy and its capability to represent certain polynomial functions, a polynomial $h(\mathbf{x})$ can be added to the right term of Eq. (1), as detailed in [21]. This extension formally complicates the system in Eq. (3), which can still be square introducing orthogonality conditions [22]. If the field to be reproduced is vectorial, Eq. (1) applies separately to each vectorial component of the field, each one treated as a scalar, leading to a number of interpolating functions equal to field dimensions [23].

RBF interpolation is particularly effective for problems involving irregularly spaced data points, such as mesh deformation [24], shape optimization [25], and aerodynamics [26], where maintaining smoothness and continuity is crucial. In the present work, RBF are at the base of the mesh morphing technique [27]: the shape of the numerical model can be changed without involving CAD by applying key displacements directly to mesh nodes. The smoothness and versatility of RBF are crucial factors in preserving the regularity of the grid, having the foresight to let buffer zones in-between the altered regions to gently absorb mesh transition. RBF enable a wide range of applications [28], many of which extend far beyond their original use in simple interpolation: image analysis [29], fast strain computation [30] and stress retrieval [31] when employed in collocation methods [32, 33], just to cite a few.

3 Response Surface Methodology

Response Surface Methodology (RSM) [34] is a statistical technique that models a desired output as an analytical function of multiple input variables, enabling efficient optimization based on extensive data collection. Typically, RSM follows a Design of Experiments (DoE) approach, which involves performing numerous numerical simulations across variations of a common baseline. These variations stem from adjustments to key parameters relevant to the study (e.g., component length, wall thickness), generating distinct input conditions for numerical analysis. By examining the resulting outputs, a mathematical function—often a polynomial [35] or an RBF [36]—is derived to represent the relationship between inputs and outputs analytically. This analytical model significantly simplifies exploration of system behavior, allowing for manual analysis or optimization via specialized software.

In the study involving the RIBES wing box, RSM was applied using RBF support. Specifically, the analytical function used to explore the design space during optimization was derived from an RBF interpolation, which established correlations between key design parameters and FEM results under consistent conditions. When a sufficient number of Design Points (DP) are strategically distributed across the design space, the RBF interpolator can predict outcome values for parameter combinations that were not explicitly tested but fall within the analyzed range. The combination of RSM and RBF has demonstrated exceptional effectiveness in exploring complex design spaces and identifying optimal solutions across various fields, including aerodynamics [37, 38], thermo-structural applications [39], and medicine [40], among others.

4 The RIBES Wing Model

The RIBES wing model was designed specifically for the aeroelastic experimental measurements campaign of the project. The tests were performed in the low-speed wind-tunnel of the University of Naples "Federico II". The facility has the characteristics listed below:

- Type: closed circuit-closed test section;
- Test section dimensions: 2.0 m × 1.4 m;
- Maximum speed: about 160 km/h (i.e., 45 m/s approximatively);
- Turbulence level: 0.1%;
- Temperature range: 10–50 °C (during test the air temp increase);
- Speed range: 5–45 m/s;
- Reynolds number: 1–2 million for airfoil 2-D tests. Usually about 0.9–1.0 million for 3D model tests (chord of about 0.25 m);
- Dynamic Pressure: 15–1200 Pa;
- Stagnation pressure: 104,700 Pa.

The model was installed on the side wall of the test section as a cantilever. It consists of a straight wing 1.6 m wide with a root chord of 600 mm and 0.7 as taper ratio. The airfoil adopted in all sections was designed starting from the Göttingen 398 scaling the original shape to a thickness t/c = 11% and redesigning the leading edge in order to improve the stall. The baseline wing box configuration was a typical aeronautical structure composed of two C-shaped spars and seven ribs. The front spar was located at 15% of the chord and kept orthogonal to the symmetry plane, whilst the rear spar was located at 70% of the chord. Figure 1 shows a sketch of the wing with main dimensions, whilst Fig. 2a, b respectively report the structure of ribs and spars supporting the wing from the interior, and a detail of the ribs-spars connection.

Table 2 lists the main geometrical dimensions of the model including the reference surface, whose value was 0.816 m^2, and the Mean Aerodynamic Chord (MAC) of 515 mm. Ribs, spars and wing skin were realized with Al 2014 T3 aluminum, whose mechanical characteristics are gathered in Table 3.

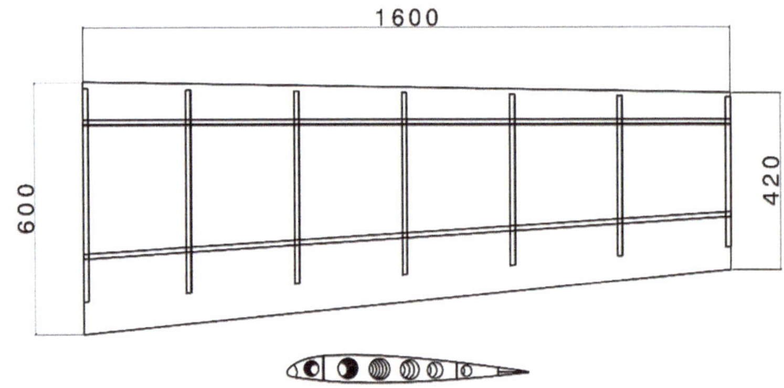

Fig. 1 Sketch of the RIBES wing with main dimensions

a b

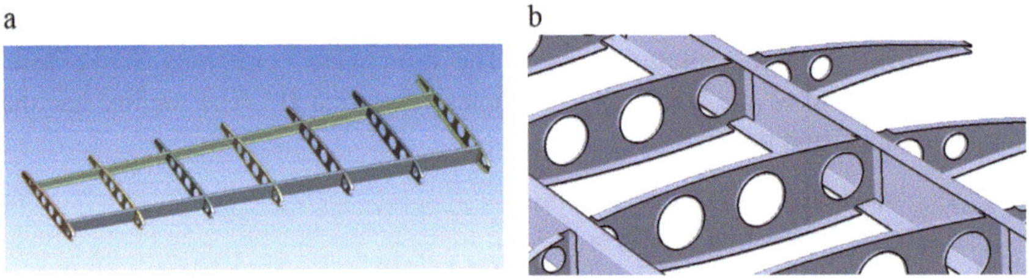

Fig. 2 Ribs and spars internal structure (**a**) and a detail of the ribs-spars connection (**b**)

Table 2 Main dimensions of the RIBES model

Geometrical parameter	Value
Span	1600 mm
Reference surface	0.816 m^2
MAC	515 mm
Root chord	600 mm
Tip chord	420 mm
Root thickness	67 mm
Tip thickness	47 mm

Table 3 Mechanical characteristics of Al 2014 T3 aluminum

Material's parameter	Value
Young's modulus	73,100 MPa
Poisson's ratio	0.33
Yield stress	290 MPa
Density	2780 kg/m^3

Fig. 3 CFD domain reproducing the wind tunnel

5 CFD Analysis

A CFD simulation was carried out to assess the aerodynamic performance of the wing and to generate an accurate load distribution required to feed the structural optimization process. The CFD domain replicated the wind tunnel test section, including the inlet convergent element, as shown in Fig. 3. The mesh was a multi-block structured grid, consisting of 3.5 million hexahedral elements. The boundary layer was solved up to the wing model's surface. The analysis was steady-state, fully turbulent compressible RANS with a Reynold's number of 1.41×10^6. The turbulence model adopted was the k-ω SST of Menter [41]. The resulting pressure distribution on the wing surfaces, illustrated in Fig. 4, was then mapped onto the CSM model as the input load for the structural analysis.

6 Optimization Setup

The FSI analysis of the wing was conducted according to the one-way coupling paradigm, which involves a sequential approach where the fluid flow and structural response are analyzed independently, with data transferred in only one direction. The optimization was then conducted on the structural model without reverting to the CFD module. The CSM structural model was built in Ansys APDL. The computational grid was realized with 85,500 elements featuring bonded contact at the

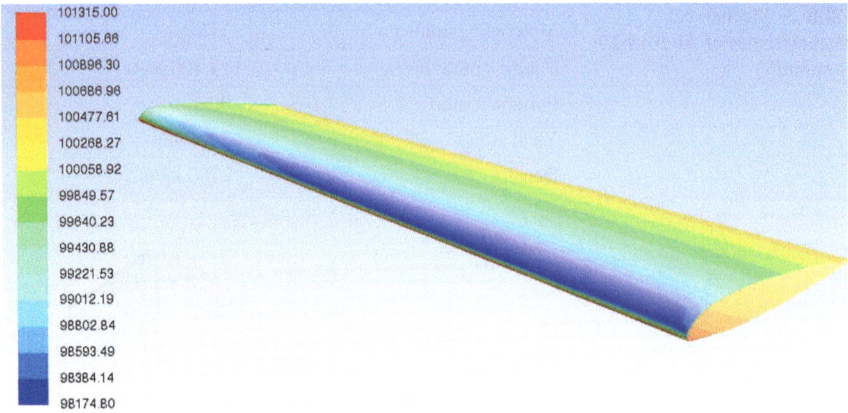

Fig. 4 CFD pressure distribution

interfaces between the box wing elements. The pressure distribution was mapped from CFD onto CSM elements directly resorting to the APDL command features, bridging the gap between mismatching meshes using the mapping tool developed within the RIBES project [12]. The combined use of RBF and fuzzy sets ensured force and moment balance between the fluid and structural domains, enabling precise and seamless data exchange across the different modules. The optimization study relied on the RBF4AERO platform, which could both modify the characteristics and apply shape modifications to the numerical model. As regards the latter, they were obtained through the RBF-based Morpher Tool (MT) integrated into the RBF4AERO platform. The exploration of the design space focused on specific modifications carefully selected to meet these criteria as effectively as possible while preserving the realistic representativeness of the mock-up. The variables of design were:

- **The position of the two ribs near the wing root** (highlighted in Fig. 5)—displacements were 30, 60 and 90 mm towards the wing root.
- **The thickness of the spars and wing skin**—the values were 0.25, 0.5, 0.75 and 1 mm (the baseline values for all components was 0.5 mm);

The two ribs were selected for their expected structural role, as it was supposed them to have significant impact on the deformation of the wing assembly. The rib positions were adjusted using mesh morphing techniques without altering the external shape of the wing. Since the one-way FSI approach does not require additional CFD analyses, the aerodynamic model remained unchanged.

Shape modifications were achieved using the MT integrated within the RBF4AERO Platform. The RBF solutions were applied using the so-called "Encap" feature (i.e., encapsulation domain): a stationary encapsulation domain was defined to restrict the morphing effect to the nodes near the rib being adjusted, while a moving encapsulation domain was created to encompass all nodes of the selected rib with

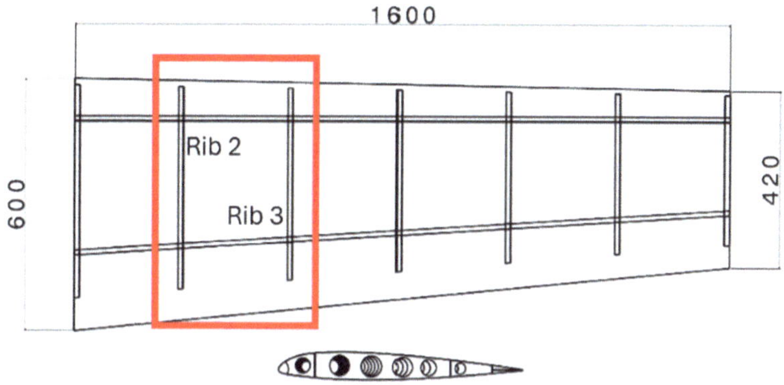

Fig. 5 Ribs selected for optimization study

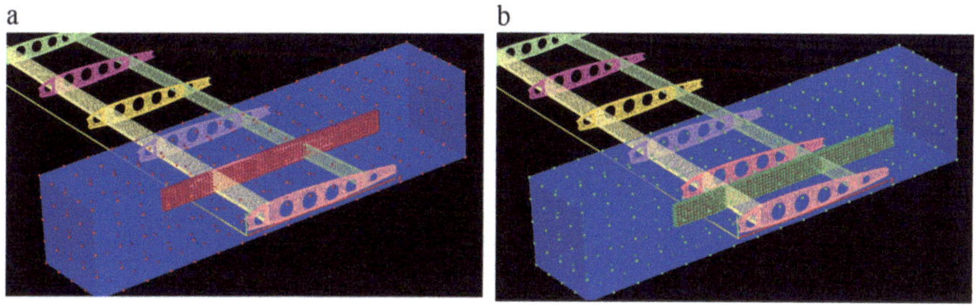

Fig. 6 Source points display—initial position (**a**), source points preview—final position (**b**)

the same Y-coordinates, as illustrated in Fig. 6. Both translation and scaling transformations were applied to each rib, thus ensuring the wing skin surface to remain unchanged.

The MT enabled the creation of various configurations without the need for remeshing. In the example shown in Fig. 7, the impact of rib displacements on the baseline configuration is clearly visible.

An optimization aimed at balancing the conflicting requirements that the RIBES wing had to meet, and considering both static and buckling effects, was conducted as a parametric study based on the DoE outlined in Table 4, for spars and skin thicknesses, and Table 5, for the displacement of Rib 2 and Rib 3. The maximum deflection and equivalent stress calculated according to the von Mises method were assigned as the objective functions.

6.1 *Results of the Study*

The effect of rib positions on the objective functions showed to be minimal. The observed variation of maximum displacement when translating the ribs was very

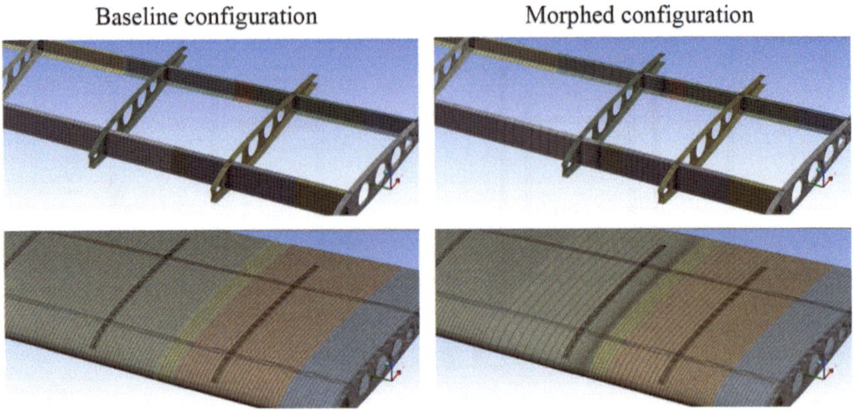

Fig. 7 Effect on ribs and spars (above) and on wing skin (below) of moving rib2 by 30 mm and rib3 by 90 mm

Table 4 Thickness variation

Design point	Spars thickness (mm)	Skin thickness (mm)
DP 0	0.50	0.50
DP 1	0.25	0.50
DP 2	0.75	0.50
DP 3	1.00	0.50
DP 4	0.50	0.75
DP 5	0.25	0.75
DP 6	0.75	0.75
DP 7	1.00	0.75
DP 8	0.50	1.00
DP 9	0.25	1.00
DP 10	0.75	1.00
DP 11	1.00	1.00

moderate, ranging from 9.77 to 9.83 mm, with a reduction of 0.61% with respect to the baseline positions. Similarly, the maximum equivalent stress showed a minor decrease, ranging from 38.027 to 37.209 MPa and reflecting a 2.14% reduction from the baseline. Figures 8 and 9 respectively show the response surfaces for the maximum displacement and maximum equivalent stress as rib positions were adjusted. The Response surfaces were obtained sampling the design space by the Latin Hypercube method [42]. This strategy places DP across the design space, more efficiently than a random positioning, and avoiding the need for more computationally expensive approaches like the full-factorial method. As for the buckling multiplier, the DoE-based optimization analysis revealed an improvement of 7.63%.

Table 5 Ribs position variation

Design point	Rib 2 displacement (mm)	Rib 3 displacement (mm)
DP 0	0	0
DP 1	0	30
DP 2	0	60
DP 3	0	90
DP 4	30	0
DP 5	30	30
DP 6	30	90
DP 7	60	0
DP 8	60	30
DP 9	60	60
DP 10	60	90
DP 11	90	30
DP 12	90	60
DP 13	90	90

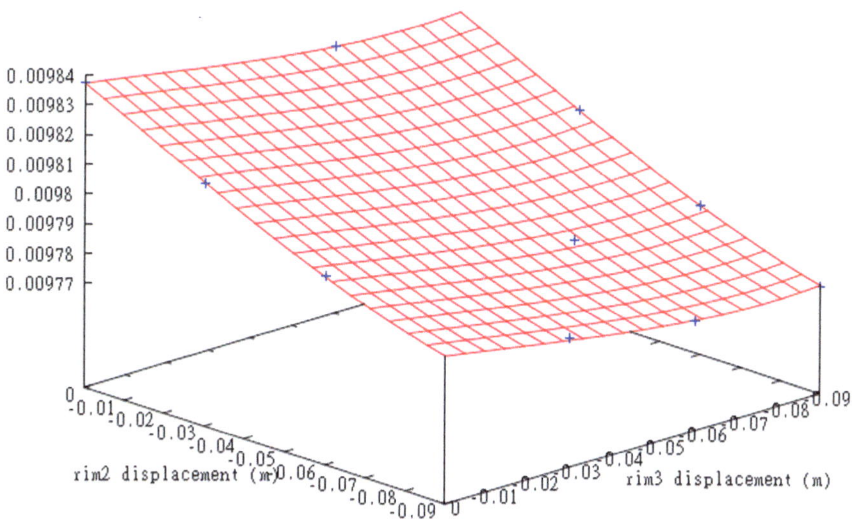

Fig. 8 Response surface of maximum displacement versus ribs positions

The thickness of the skin and spars had a significantly greater impact on both maximum displacement and equivalent stress. Maximum displacement varied between 4.95 and 9.74 mm with a reduction of up to 49.18% from the baseline configuration, whilst the maximum equivalent stress ranged from 25.684 to 37.154 MPa showing a maximum decrease of 30.87%. Figures 10 and 11 respectively display the

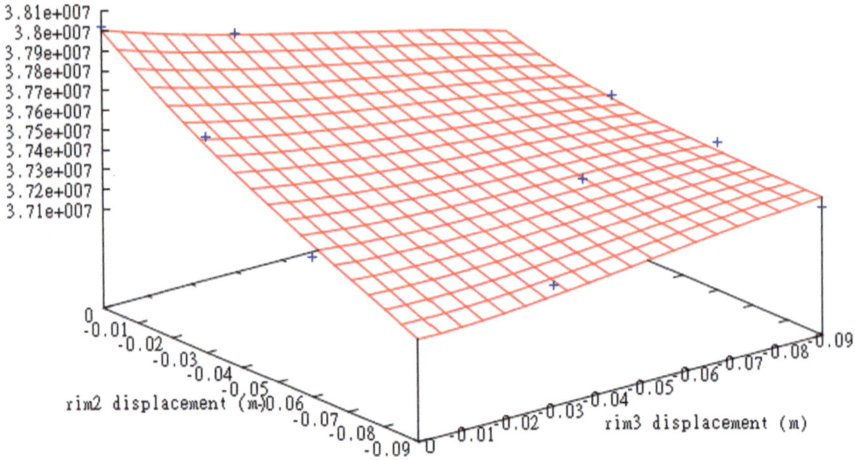

Fig. 9 Maximum equivalent stress versus ribs positions

response surfaces for maximum displacement and equivalent stress as skin and spar thicknesses were tweaked.

The DoE-based optimization analysis further demonstrated that selecting different thickness combinations, an increase of the buckling load multiplier by two orders of magnitude could be gained.

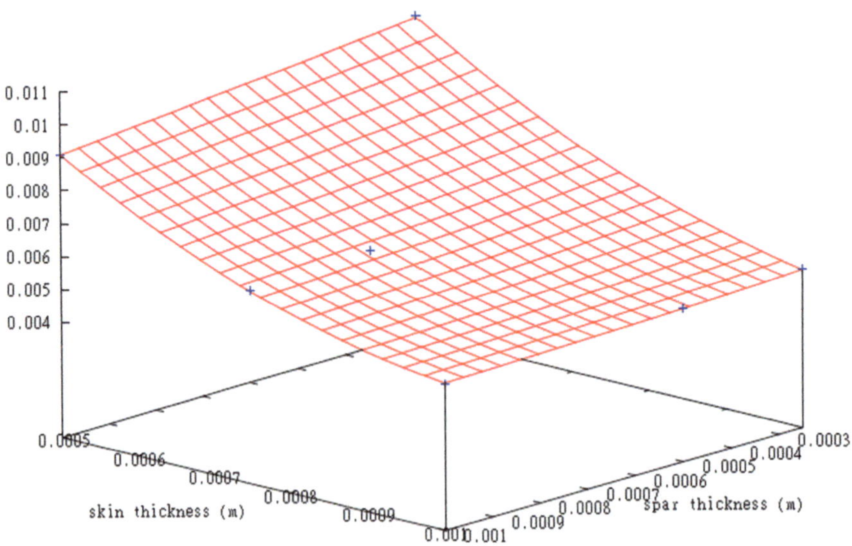

Fig. 10 Response surface of maximum displacement versus skin and spar thicknesses

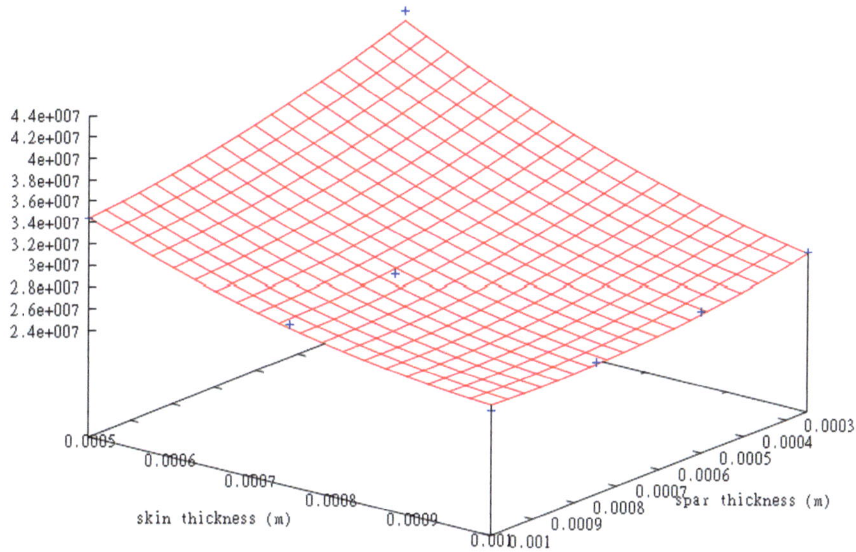

Fig. 11 Response surface of maximum equivalent stress versus skin and spar thicknesses

7 Conclusions

This paper focused on the structural optimization of the RIBES wind tunnel model consisting in a wing replicating a realistic wing box topology. The tools adopted for the study are based on the numerical platform developed within the EU RBF4AERO project. The analysis of the model, designed for an aeroelastic measurement campaign, was based on the one-way FSI paradigm with a unique data transfer from CFD to CSM model. The structural analysis was the scenario for an optimization environment based on a parametric study in which the space of design variables was populated with the position of two ribs and with the thickness of both skin and spars. The shape modification was implemented by a mesh morphing tool based on Radial Basis Functions. This approach enables efficient and smooth transformations of numerical domains without the need for remeshing. The results of the optimization, explored by Response Surfaces generated adopting RBF based interpolation, showed the thickness of spars and skin to play a much more effective role, with respect to the adjustment of ribs position, in reducing wing deformations, minimizing stresses while improving the buckling load capacity.

References

1. Corrado, G., Ntourmas, G., Sferza, M., Traiforos, N., Arteiro, A., Brown, L., et al.: Recent progress, challenges and outlook for multidisciplinary structural optimization of aircraft and aerial vehicles. Prog. Aerosp. Sci. **135**, 100861 (2022)
2. Zuo, Y., Chen, P., Fu, L., Gao, Z., Chen, G.: Advanced aerostructural optimization techniques for aircraft design. Math. Probl. Eng. **2015**(1), 753042 (2015)
3. Bhatia, M., Kapania, R.K., Haftka, R.T.: Structural and aeroelastic characteristics of truss-braced wings: a parametric study. J. Aircr. **49**(1), 302–310 (2012)
4. Kennedy, G.J., Martins, J.R.: A parallel aerostructural optimization framework for aircraft design studies. Struct. Multidiscip. Optim. **50**, 1079–1101 (2014)
5. Benra, F.K., Dohmen, H.J., Pei, J., Schuster, S., Wan, B.: A comparison of one-way and two-way coupling methods for numerical analysis of fluid-structure interactions. J. Appl. Math. **2011**(1), 853560 (2011)
6. Wang, T., Wüchner, R., Sicklinger, S., Bletzinger, K.U.: Assessment and improvement of mapping algorithms for non-matching meshes and geometries in computational FSI. Comput. Mech. **57**, 793–816 (2016)
7. Kang, H.S., Kim, Y.J.: Optimal design of impeller for centrifugal compressor under the influence of one-way fluid-structure interaction. J. Mech. Sci. Technol. **30**, 3953–3959 (2016)
8. Marzec, Ł., Buliński, Z., Krysiński, T., Tumidajski, J.: Structural optimisation of H-rotor wind turbine blade based on one-way fluid structure interaction approach. Renew. Energy **216**, 118957 (2023)
9. RIBES: http://RIBES-project.eu/project. Accessed: 2024-09-09
10. Biancolini, M.E., Cella, U., Groth, C., Chiappa, A., Giorgetti, F., Nicolosi, F.: Progresses in fluid-structure interaction and structural optimization numerical tools within the EU CS RIBES project. In: Evolutionary and Deterministic Methods for Design Optimization and Control with Applications to Industrial and Societal Problems, pp. 529–544 (2019)
11. Cella, U., Biancolini, M.E., Groth, C., Chiappa, A., Beltramme, D.: Development and validation of numerical tools for FSI analysis and structural optimization: the EU RIBES project status. In: AIAS 44th National Congress, vol. 2, no. 5 (2015)
12. Biancolini, M.E., Chiappa, A., Giorgetti, F., Groth, C., Cella, U., Salvini, P.: A balanced load mapping method based on radial basis functions and fuzzy sets. Int. J. Numer. Methods Eng. **115**(12), 1411–1429 (2018)
13. Groth, C., Porziani, S., Chiappa, A., Giorgetti, F., Cella, U., Nicolosi, F., et al.: Structural validation of a realistic wing structure: the RIBES test article. Procedia Struct. Integr. **12**, 448–456 (2018)
14. Cella, U., Della Vecchia, P., Groth, C., Porziani, S., Chiappa, A., Giorgetti, F., et al.: Wind tunnel model design and aeroelastic measurements of the RIBES wing. J. Aerosp. Eng. **34**(1), 04020109 (2021)
15. https://www.rbflab.eu/projects/102_rbf4aero/. Accessed 2024-10-15
16. RBF4AERO: https://cordis.europa.eu/project/rcn/109141/factsheet/en. Accessed: 2024-09-09
17. Kapsoulis, D.H., Asouti, V.G., Giannakoglou, K.C., Porziani, S., Costa, E., Groth, C., et al.: Evolutionary aerodynamic shape optimization through the RBF4AERO platform. In: ECCOMAS Congress 2016, VII European Congress on Computational Methods in Applied Sciences and Engineering, pp. 5–10 (2016)
18. Biancolini, M.E., Cella, U.: Flexible Engineering Toward Green Aircraft. Springer eBooks (2020). https://doi.org/10.1007/978-3-030-36514-1
19. Davis, P.J.: Interpolation and Approximation. Courier Corporation (1975)
20. Fasshauer, G.E.: Meshfree Approximation Methods with MATLAB. World Scientific Pub Co Inc. (2007)
21. Boyd, J.P., Gildersleeve, K.W.: Numerical experiments on the condition number of the interpolation matrices for radial basis functions. Appl. Numer. Math. **61**(4), 443–459 (2011)
22. Buhmann, M.D.: Radial basis functions. Acta Numer. **9**, 1–38 (2000)

23. Cella, U., Biancolini, M.E.: Aeroelastic analysis of aircraft wind-tunnel model coupling structural and fluid dynamic codes. J. Aircr. **49**(2), 407–414 (2012)
24. De Boer, A., Van der Schoot, M.S., Bijl, H.: Mesh deformation based on radial basis function interpolation. Comput. Struct. **85**(11–14), 784–795 (2007)
25. Groth, C., Chiappa, A., Biancolini, M.E.: Shape optimization using structural adjoint and RBF mesh morphing. Procedia Struct. Integr. **8**, 379–389 (2018)
26. Biancolini, M.E., Costa, E., Cella, U., Groth, C., Veble, G., Andrejašič, M.: Glider fuselage-wing junction optimization using CFD and RBF mesh morphing. Aircr. Eng. Aerosp. Technol. **88**(6), 740–752 (2016)
27. Biancolini, M.E., Viola, I.M., Riotte, M.: Sails trim optimisation using CFD and RBF mesh morphing. Comput. Fluids **93**, 46–60 (2014)
28. Biancolini, M.E.: Fast Radial Basis Functions for Engineering Applications, pp. 1–36. Springer International Publishing, Cham (2017)
29. Groth, C., Chiappa, A., Porziani, S., Biancolini, M.E., Marotta, E., Salvini, P.: A post-processing method based on radial basis functions for the fast retrieval of the strain field in digital image correlation methods. Materials **15**(22), 7936 (2022)
30. Groth, C., Chiappa, A., Porziani, S., Salvini, P., Biancolini, M.E.: An RBF meshless approach to evaluate strain due to large displacements in flexible printed circuit boards. Micromachines **13**(8), 1163 (2022)
31. Chiappa, A., Groth, C., Biancolini, M.E.: A two-scale RBF meshless method for the interface stress retrieval in simply bended and torqued long-fibres laminates. Compos. Struct. **306**, 116600 (2023)
32. Kansa, E.J.: Multiquadrics—a scattered data approximation scheme with applications to computational fluid-dynamics-I surface approximations and partial derivative estimates. Comput. Math. Appl. **19**(8–9), 127–145 (1990)
33. Kansa, E.J.: Multiquadrics—a scattered data approximation scheme with applications to computational fluid-dynamics-II solutions to parabolic, hyperbolic and elliptic partial differential equations. Comput. Math. Appl. **19**(8–9), 147–161 (1990)
34. Khuri, A.I., Mukhopadhyay, S.: Response surface methodology. Wiley Interdiscip. Rev. Comput. Stat. **2**(2), 128–149 (2010)
35. Hosder, S., Watson, L.T., Grossman, B., Mason, W.H., Kim, H., Haftka, R.T., Cox, S.E.: Polynomial response surface approximations for the multidisciplinary design optimization of a high speed civil transport. Optim. Eng. **2**, 431–452 (2001)
36. McDonald, D.B., Grantham, W.J., Tabor, W.L., Murphy, M.J.: Global and local optimization using radial basis function response surface models. Appl. Math. Model. **31**(10), 2095–2110 (2007)
37. Deng, K., Chen, H.: A hybrid aerodynamic optimization algorithm based on differential evolution and RBF response surface. In: 17th AIAA/ISSMO Multidisciplinary Analysis and Optimization Conference, p. 3671 (2016)
38. Di Meo, E., Lopez, A., Groth, C., Biancolini, M.E., Valentini, P.P.: Reduced-order model of a time-trial cyclist helmet for aerodynamic optimization through mesh morphing and enhanced with real-time interactive visualization. Fluids **9**(12), 300 (2024)
39. Groth, C., Chiappa, A., Guarino, R.: An analytical thermal model for the optimization of EU DEMO feeder thermal shields. Fusion Eng. Des. **207**, 114650 (2024)
40. Kardampiki, E., Vignali, E., Haxhiademi, D., Federici, D., Ferrante, E., Porziani, S., et al.: The hemodynamic effect of modified Blalock–Taussig shunt morphologies: a computational analysis based on reduced order modeling. Electronics **11**(13), 1930 (2022)
41. Menter, F.R.: Two-equation eddy-viscosity turbulence models for engineering applications. AIAA J. **32**(8), 1598–1605 (1994)
42. Iman, R.L., Davenport, J.M., Zeigler, D.K.: Latin Hypercube Sampling (Program User's Guide) [LHC, in FORTRAN] (No. SAND-79-1473). Sandia Labs, Albuquerque, NM (USA) (1980)

A Mixed Intrusive and Non-intrusive Approximation Technique for Efficient Robust Aerodynamic Shape Design

Domenico Quagliarella

Abstract Initially conceived in financial engineering, advanced risk functions, such as Value-at-Risk (VaR) and Conditional Value-at-Risk (CVaR), play an important role in robust optimization problems of engineering interest. The focus is here on techniques for increasing the computational efficiency of robust optimization processes based on risk functions, whose computation is computationally demanding. In previous works, we illustrated an intrusive approach based on gradient computation and another non-intrusive approach based on Gaussian processes. Here, we analyze the performance of a mixed approach, where both the gradient information, calculated intrusively, and the approximation methodology based on Gaussian processes are used simultaneously. The new mixed approximation technique is compared and discussed in terms of efficiency and easiness of implementation differences compared to the intrusive and non-intrusive ones.

Keywords Engineering design · Evolutionary optimization · Risk measures · Robust optimization · Gaussian processes · Adjoint-based gradient computation

1 Introduction

The use of robust design methods reveals its great potential in design optimization applications in which it is not possible to take into account in the design phase all the parameters that characterize the functioning of the system in the real world.

For example, in the case of aerodynamic design, the aircraft will rarely perform strictly under the design conditions due to unpredictable variations in environmental and operational conditions. Consider the design lift coefficient (c_L) of a wing. The actual c_l at which the aircraft will operate will depend on the payload and the fuel level, which varies significantly during the mission, and the weather conditions affect its actual speed. In addition, the actual shape of the wing can differ significantly from the theoretical design both for machining tolerances and for deformations induced by

D. Quagliarella (✉)
Department of Fluid Mechanics, Italian Aerospace Research Centre (CIRA), Capua, Italy
e-mail: d.quagliarella@cira.it

M. E. Biancolini et al. (eds.), *Digital Twins for Aerospace*, Springer Aerospace Technology, https://doi.org/10.1007/978-3-032-05022-9_3

the aerodynamic load and, finally, it must be considered that the predictive numerical models are often characterized by epistemic uncertainty or by limits in the ability to model the real world. In the robust approach to design optimization, therefore, the quantities of interest (QoI) being optimized are assimilated to random variables whose distribution depends on a certain number of input variables, which are, in turn, modeled as random variables with a known distribution.

The design process requires optimizing a quantity of interest known only in a statistical sense and which must be measured with statistical risk functions [12]. The most commonly used risk functions in the literature are the mean and standard deviation, and some authors also use higher-order statistical moments. In this work, on the other hand, more sophisticated risk functions are used, such as the Value-at-Risk (VaR) and the Conditional Value-at-Risk (CVaR) [10, 15], which, among the various advantages, present one that in our opinion is substantial, that is the ability to describe the behavior of the statistical distribution tails with greater accuracy. Furthermore, another significant advantage is the asymmetry of these functions. In fact, while both positive and negative events influence mean and variance, VaR and CVaR make it possible to give greater importance to adverse events. The disadvantage of the robust design approach via risk functions is linked to the increase, often more than an order of magnitude, in the number of evaluations of the QoI. Often, the use of sophisticated risk functions such as CVaR requires a denser sampling of the distribution tails, which results in an even more significant increase in the computational effort required. In this context, techniques to reduce the computational weight of the computation of the statistical parameters of the QoI distribution gain importance. In the literature, tools such as stochastic collocation [18] or polynomial chaos [16] are widespread and are often used to build surrogate models of the random distribution to be characterized. These methods can be non-intrusive, i.e., based on a simple sampling of the quantity of interest, or intrusive, i.e., based on quantities that require profound access to the computation method used, as the computation of the gradient of the QoI. Indeed, efficient gradient computation may require very sophisticated and problem-dependent methods, like the numerical solution of the adjoint field. This work describes a mixed intermediate approach that uses both intrusively derived information, particularly the gradient of the nominal configuration, and a classical, non-intrusive coarse Monte Carlo sampling. The information thus obtained is used to build an approximation of the distribution function through a Gaussian process. It is worth noting that we consider a method intrusive when it requires additional physical information not readily available from the numerical model used to calculate the QoI. In the case study presented here, the QoI is the drag coefficient of an airfoil that we obtain using a Reynolds-averaged Navier–Stokes (RANS) equations solver, and the additional information is the gradient of the QoI obtained from the solution of the adjoint field of the RANS equations. Therefore, even if the adjoint field solver is already available, it uses field information coming from the RANS solver, and this is, according to our definition, an intrusive process since it re-elaborates physical data coming from the RANS solver.

The work is organized as follows. We will not discuss the advantages and disadvantages of VaR and CVaR in detail because this has been done in other works [9].

However, for the sake of completeness, we will report the fundamental elements of the intrusive and non-intrusive approach we used, referring to the implementation details with appropriate references. Then, we will illustrate the hybrid approach that mixes information on the cumulative distribution obtained with intrusive and non-intrusive methods.

The data on which we will show the application of the method relate to the same aerodynamic design problem presented for the illustration of the intrusive [7] and non-intrusive methods [8]. We will, therefore, only briefly summarize the application problem and refer to the two aforementioned publications for details. Furthermore, since the focus is on the hybrid approximation method of the cumulative distribution function and the VaR and CVaR risk functions, we will not repeat the execution of the robust optimization problem, but we will simply recalculate the distributions with the new approximation method.

2 Robust Approach to Design Optimization

For the sake of clarity, we provide a brief introduction to the problem of robust optimization and the use of advanced risk functions. In real-world conditions, an optimal solution obtained through an optimization procedure often performs worse than expected due to uncertainties that were not foreseen in the design phase. A possible solution to this problem is to model the uncertain problem parameters with appropriate statistical distributions and to sample the value of the Quantity of Interest (QoI) to be optimized to estimate its actual behavior in the real world [12]. In this way, the QoI no longer assumes a single value at a particular instance of the design parameters but is a statistical distribution. The distribution of the QoI depends on the design variables, and the problem arises of evaluating how much of one distribution is better than another. Classical statistical moments, such as mean and variance or a combination, are often used for this purpose. The drawback of these classical approaches is that they penalize all distributions with many values far from the mean value unless high-order moments are used. Instead, in many robust optimization problems, a distribution evaluation function that works asymmetrically and penalizes only the configurations that produce a worsening of the QoI is better suited. In particular, the Value-at-Risk [6] (VaR) and the Conditional Value-at-Risk [14, 15] (CVaR) are statistical measures that exhibit this asymmetry [9]. They were first used in the financial sector to evaluate the market risk or credit risk of a portfolio and can be directly derived from the Cumulative Distribution Function (CFD). The CDF gives the area under the probability density function from minus infinity to x so that $F_X(x) = P(X \leq x)$ is the probability that X takes on a value $\leq x$. However, the number of QoI samples is necessarily finite in the applications illustrated here. Therefore, we will use an approximation of the cumulative distribution, namely the Empirical Cumulative Distribution Function (ECDF) [1, 17].

3 Algorithm Description

The core algorithm is a Gaussian Process that starts from two different cumulative distributions: an approximation built from the gradient calculated in the nominal condition and a relatively coarse Monte Carlo (MC) sampling. The approximation of the risk functions thus obtained is better than that obtained with the gradient alone or with the Gaussian Process constructed only by taking some elements of the MC sampling. Below, we describe the various components of the procedure.

The optimization process involves iterating through two distinct phases: training the surrogate model to approximate the ECDF and a subsequent robust optimization phase of the QoI using the previously obtained surrogate model. This cycle can be repeated an arbitrary number of times by adding to the training set some of the ECFDs computed starting from the solution of the robust optimization problem. This iterative process aims to increasingly refine and improve the surrogate model of the ECDF in proximity to the optimum.

Due to the highly combinatorial nature of the surrogate ECDF training process and the relatively short time required to evaluate the objective function, a simple genetic algorithm is used [4]. On the other hand, the obtained surrogate model is parametric and involves the evaluation of a few true QoIs to work correctly, so a more efficient optimization algorithm is needed. In the reported example, the Covariance Matrix Adaptation Evolution Strategy (CMA-ES) [5] is used for this purpose.

This last point should not be surprising because estimating an ECDF with a Monte Carlo sampling typically requires a very high number of samples that can even be, for the type of applications we are interested in, of the order of hundreds. The surrogate model we propose requires a relatively small number of samples in the order of 4–6. However, we must add those used to calculate the ECDF with adequate samples for the training phase to the total number of QoI evaluations.

3.1 Gradient-Based ECDF and Risk Function Approximation

The technique is described and validated in detail in [7]. In summary, the algorithmic strategy is as follows. An intrusive approach is used to efficiently compute the gradients of the QoI w.r.t the uncertain variables; then, a linear approximation of the QoI is built by means of the extracted gradients:

$$q(\mathbf{z}) \approx q(\mathbf{z}_0) + \sum_{i=1}^{n} \frac{\partial q(\mathbf{z}_0)}{\partial z^{(i)}} \left(z^{(i)} - z_0^{(i)} \right) \tag{1}$$

where $\mathbf{z}$ is the vector of the uncertain variables modeled as random distributions.

This allows the calculation of the approximated Empirical Cumulative Distribution Function (ECDF) used for CVaR estimation. In the applications considered here, the QoI is the drag coefficient (c_d) of an airfoil calculated through the Reynolds Averaged Navier–Stokes (RANS) equations, and the gradient is calculated by solving the adjoint field of the RANS at roughly the cost of one RANS flow solution.

3.2 Gaussian Process-Based Approximation

Also in this case, see [8] for an in-depth description of the method and its validation in an aerodynamic design problem. Following a pattern similar to the gradient-based ECDF approximation, the regression method based on Gaussian processes is used to derive the ECDF approximation for the QoI. After that, the statistics of interest, such as mean and standard deviation, are calculated from the approximation of the ECDF obtained with the GPs. The approach based on Gaussian processes is here implemented in a straightforward way, without resorting to sophisticated techniques such as sparse Gaussian processes or adaptive sampling.

Figure 1 briefly illustrates the core of the approximation process. We start from a Monte Carlo sampled ECDF with a medium-high number of samples (the actual number depends on the uncertain parameter set size). A small number of samples are extracted from this ECDF, which is then used to construct a Gaussian process that approximates the entire distribution. Obviously, with the same GP training parameters, the overall quality of the approximation depends on which samples are chosen. It is possible to estimate this quality by measuring the distance (with a simple Euclidean norm) between the ECDF obtained with the GP and that relating to the MC sampling.

Then it is possible to optimize the choice of the subset of samples to minimize the distance between ECDF GP and ECDF MC. Once the size of the sample subset has been established, it is a typical combinatorial optimization problem that we address here using a simple binary-encoded genetic algorithm.

Once the optimal subset is identified, the approximation of a new ECDF is obtained simply by evaluating the QoI in correspondence with the input parameters of the samples thus identified. Therefore, if we can get a good approximation with a subset of m

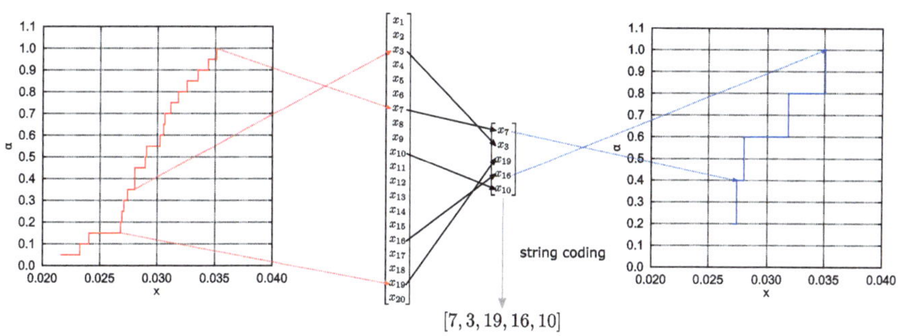

Fig. 1 Algorithm scheme of the GP approximator core

elements (with $m << n$, where n is the size of the MC sampling), the computational cost reduction obtained is equal to n/m (excluding the time required for training the GP, generally negligible compared to the cost of a single QoI). Of course, if training is done using only one ECDF, the approximation quality is likely to deteriorate when moving away from it. For this reason, it is advisable to train not on a single ECDF but starting from an appropriate training set. In [8], an iterative approach was chosen, which consists of alternating the training phases of the GP with those of robust QoI optimization to increase the quality of the training set progressively.

However, this measure alone may not be enough. Indeed, CVaR is not always well approximated due to slight differences in the upper tail of the distribution. Therefore, we introduce in the objective function a further term that accounts for the minimization of the difference between the CVaRs computed with MC sampling and the ones calculated with GPs ($\overline{\text{CVaR}}$). Still, a small error in CVaR estimation with GP can lead to the overturning of the order relationship: e.g., Monte Carlo sampling indicates $\text{CVaR}_1 < \text{CVaR}_2$ and the GP approximation $\overline{\text{CVaR}}_1 > \overline{\text{CVaR}}_2$. Therefore we introduced a further penalty term computed consider the set of pairs (CVaR, $\overline{\text{CVaR}}$) and reordering them so that $\text{CVaR}_i \leq \text{CVaR}_{i-1}$

$$P_{\text{tset}} = w \sum_{i=2}^{n} \mathbb{1}_C \left(\overline{\text{CVaR}}_i > \overline{\text{CVaR}}_{i-1} \right) \tag{2}$$

$$\mathbb{1}_C(x) := \begin{cases} 1 & \text{if } x \in C, \\ 0 & \text{if } x \notin C, \end{cases} \tag{3}$$

3.3 *Hybrid Algorithm*

The motivation that prompted us to hybridize the two previous approaches is the prospect of further reducing the subset of samples needed to obtain a good quality GP approximation.

The basic idea is to build the training subset by mixing values obtained from MC sampling with those coming from the gradient-based approximation (GBA). However, as can also be seen from the observation of Fig. 2, it may likely happen that the ECDF GBA can be very different from the ECDF MC.

Again, the remedy is to implement an appropriate combinatorial optimization process. The core of this process is shown in Fig. 3. Unlike the previous case, the majority of samples are chosen from the approximate distribution, while the MC distribution is used to pick a small control subset. The GP approximation is constructed using all the samples thus obtained in the same way, and the combinatorial optimization process, again based on a binary-coded genetic algorithm, is used to minimize the distance between ECDF GRAD + GP and ECDF MC.

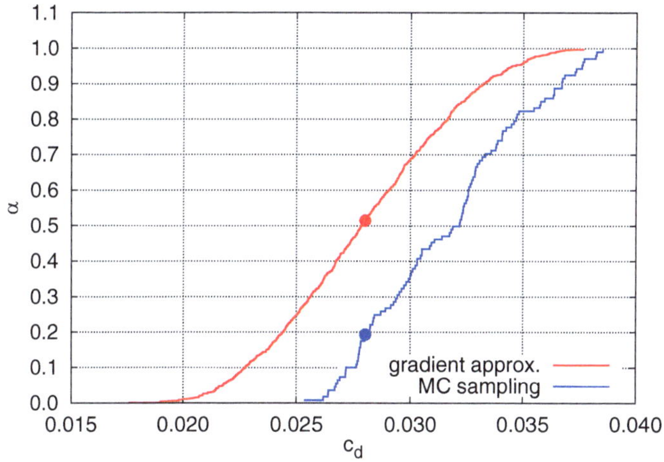

Fig. 2 ECDF differences between MC sampling and gradient-based approximation

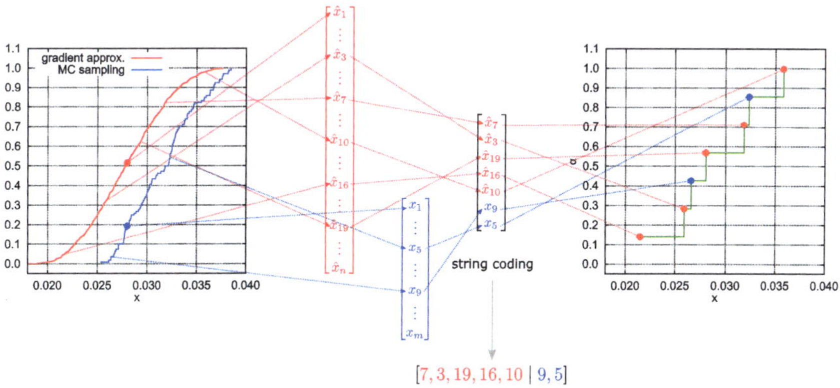

Fig. 3 Algorithm scheme of the mixed gradient + GP approximator core

4 Hybrid Approximator Training Examples

The MC ECDFs used in the following validation examples of the approach proposed here were obtained from the data relating to the aerodynamic optimization problem of a transonic airfoil described in [7, 8]. The particular definition of the robust optimization problem is not strictly necessary for understanding the examples, but it is available in detail in the references cited. Nevertheless, for the sake of clarity and completeness, we briefly describe the aerodynamic shape design problem below. The goal is the robust aerodynamic design of a Blended Wing Body (BWB) aircraft central section at the following operating conditions: Mach number = 0.8, Reynolds number = 173.52 × 106, angle of attack $\alpha = -2.86°$, lift coefficient $c_l = 0.10$. The airfoil geometric characteristics are: Maximum thickness $t/c_{\max} = 0.16$, Leading

Edge Radius (LER) = 0.0156, Trailing Edge Angle (TEA) = 27.57°, thickness t/c_{85} at 85% chord (TAT) = 0.083.

The robust design problem accounts for the geometric uncertainties related to the wing section shape and the uncertainties associated with the actual operating conditions. A uniform random perturbation U_j represents the uncertainty of the wing shape, and the uncertainties in the operational parameters Mach and AoA are modeled as four-parameter beta distributions.

The lift coefficient constraint is satisfied explicitly by changing the angle of attack, while the maximum airfoil thickness to chord ratio is satisfied by adequately scaling the airfoil shape. All the other constraints are quadratic penalties, and the robust design problem is formulated as:

$$\min_{\mathbf{w}\in W\subseteq\mathbb{R}^n} \text{CVaR}^{0.9}\left(C_d(\mathbf{w},\mathbf{z})\right) + P(\mathbf{w}) \tag{4}$$

with

$$\begin{aligned} P(\mathbf{w}) = {} & k^1 p^+(\text{LER}, 0.00781) \\ & + k^2 p^+(\text{TEA}, 22.0°) + k^3 p^+(\text{TAT}, 0.06658) \\ & + k^4 p^+(C_m, 0.01) + k^4 p^-(C_m, 0.02) + k^5 p^+(\text{error}, 0) \end{aligned}$$

$$p^+(x,y) = \begin{cases} 0 & \text{if } x \geq y \\ (x-y)^2 & \text{if } x < y \end{cases} \quad \text{and } p^-(x,y) = \begin{cases} (x-y)^2 & \text{if } x \geq y \\ 0 & \text{if } x < y \end{cases}$$

The baseline configuration is a deterministically optimized airfoil with a constraint on the pitching stability behavior. The QoI depends on 14 uncertain parameters, and 20 design variables govern the airfoil shape. The airfoil construction, the mesh generation, and the RANS analysis necessary for the calculation of the QoI (e.g., c_d) are performed using an open source framework [11] that relies on GMSH unstructured grid generator [3] and SU2 flow solver [2].

Concerning the training phases, the genetic algorithm is the same for each of the iterations reported and is characterized by a one-point binary crossover with 80% activation probability and a binary adaptive mutation operator described in [13]. All runs are characterized by a population of 240 elements that can evolve for a maximum of 500 generations. However, the number of generations actually performed can be lower if the optimization algorithm does not detect improvements in the objective function for a given number of generations. Therefore, the training phase of STEP 1 GP stopped slightly before the maximum number of generations, i.e., at generation no. 458; also, the training of STEP 2 stopped slightly before, i.e., at generation 484.

In the robust optimization phases, CMA-ES was allowed a maximum of 801 evaluations, while the population size was set to eight individuals, and the initial standard deviation was set to 0.02.

Table 1 STEP 1—single ECDF training set

Case label	ECDF distance	GP + GRAD app.		MC sampling	
		VaR*	CVaR*	VaR	CVaR
$ECDF_1$	2.50E−04	0.03629	0.03765	0.03721	0.03805
$ECDF_2$	9.33E−04	0.03417	0.03537	0.03495	0.03785
$ECDF_3$	8.46E−04	0.03254	0.03361	0.03492	0.03821

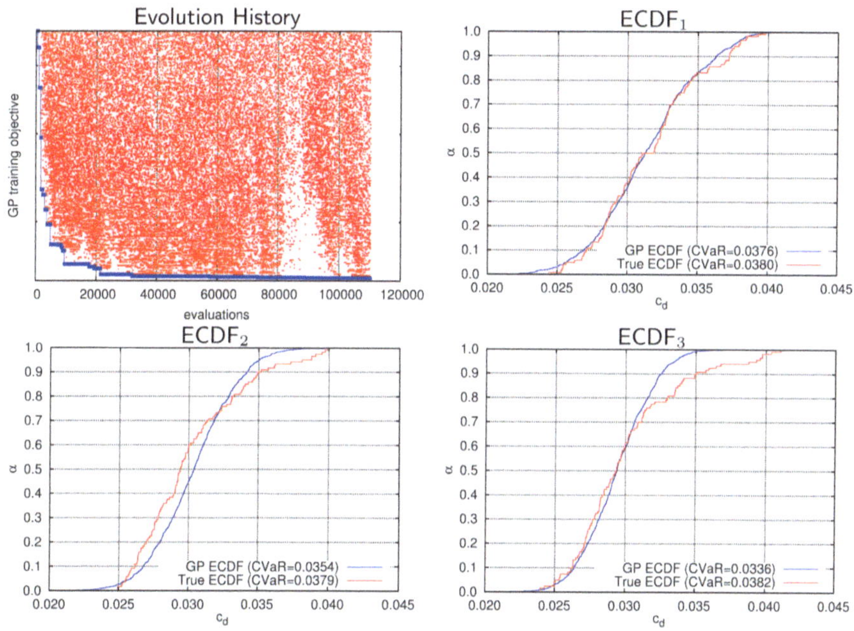

Fig. 4 STEP 1—single ECDF training set evolution history and approximation results

4.1 STEP 1—Single ECDF Training Set

This optimization run is characterized by using only the $ECDF_1$ for training. In particular, 20 points are selected from the gradient-based (GB) ECDF and 5 from the MC ECDF. Recall that the computational cost to evaluate all the GB ECDF points is approximately equal to that of only two QoI thanks to using the adjoint field for gradient evaluation. Only 20 points were chosen from the curve mentioned above in order not to make the training phase of the GP too heavy. We also remind that during the GP training phase, the QoIs are never re-evaluated because a pre-calculated ECDF database is used.

The results reported in Table 1 show how the precision of the approximation degrades considerably for the $ECDF_2$ and $ECDF_3$. This is also clearly shown in Fig. 4, where the poor approximation of $ECDF_2$ and $ECDF_3$, which are not part of the training set, is evident.

Table 2 STEP 2—three ECDF training set

Case label	ECDF distance	GP + GRAD app.		MC sampling	
		VaR*	CVaR*	VaR	CVaR
$ECDF_1$	4.38E−04	0.03625	0.03805	0.03721	0.03805
$ECDF_2$	9.25E−04	0.03598	0.03775	0.03495	0.03785
$ECDF_3$	1.18E−03	0.03617	0.03820	0.03492	0.03821

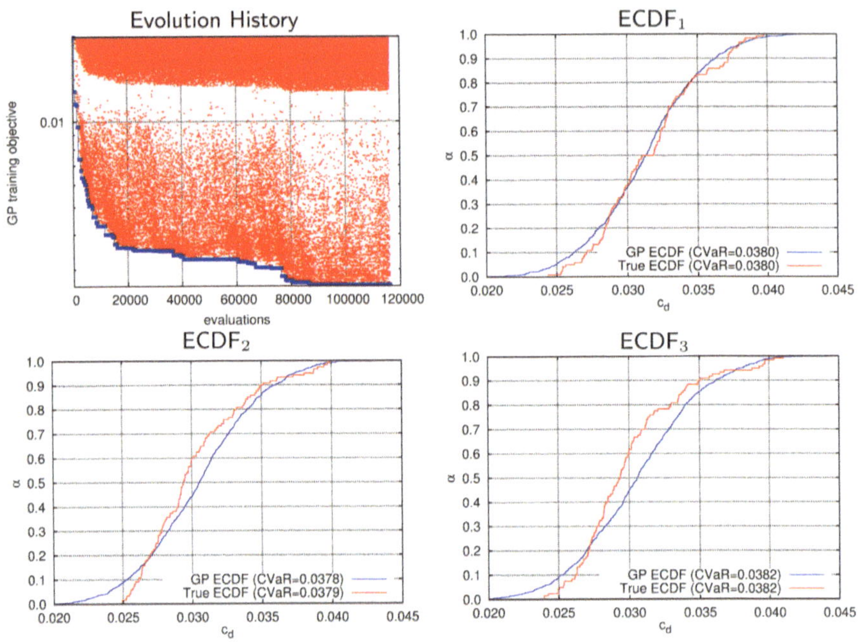

Fig. 5 STEP 2—three ECDF training set evolution history and ECDF approximation results

4.2 STEP 2—Three ECDF Training Set

To remedy the reported lack of accuracy, we made a new optimization in which we also included the $ECDF_2$ and $ECDF_3$ in the training set. The results reported in Table 2 and Fig. 5 show that, this time, the approximation precision does not degrade and that the order relationship, which is of primary importance in optimization, is also maintained.

Obviously, it is reasonable to expect a degradation of precision as we move away from the training set, and that is why, in our opinion, the procedure should be applied iteratively and incrementally, as shown in [8].

Table 3 STEP 3—reduced number of samples from MC ECDF

Case label	ECDF distance	GP + GRAD app.		MC sampling	
		VaR*	CVaR*	VaR	CVaR
$ECDF_1$	6.13E−04	0.03598	0.03748	0.03721	0.03805
$ECDF_2$	1.22E−03	0.03639	0.03793	0.03495	0.03785
$ECDF_3$	1.57E−03	0.03625	0.03781	0.03492	0.03821

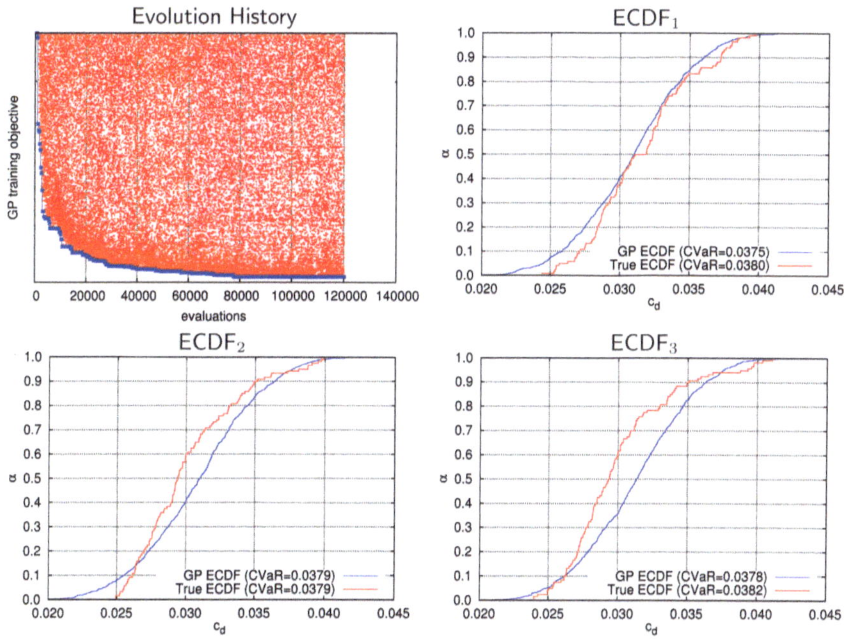

Fig. 6 STEP 3—reduced number of samples from MC ECDF test case—evolution history and approximation results

4.3 STEP 3—Reduced Number of Samples from MC ECDF

In this case, we kept all three ECDFs in the training set, but we tried to select just one point from each MC ECDF. As seen from Table 3 and Fig. 6, the approximation results, even if characterized by a small error, are not suitable for optimization because the CVaRs calculated with the GP ECDFs do not always respect the order relationship.

Table 4 STEP 4—two points from MC ECDFs

Case label	ECDF distance	GP + GRAD app.		MC sampling	
		VaR*	CVaR*	VaR	CVaR
$ECDF_1$	4.96E−04	0.03630	0.03815	0.03721	0.03805
$ECDF_2$	1.12E−03	0.03634	0.03807	0.03495	0.03785
$ECDF_3$	1.38E−03	0.03641	0.03826	0.03492	0.03821

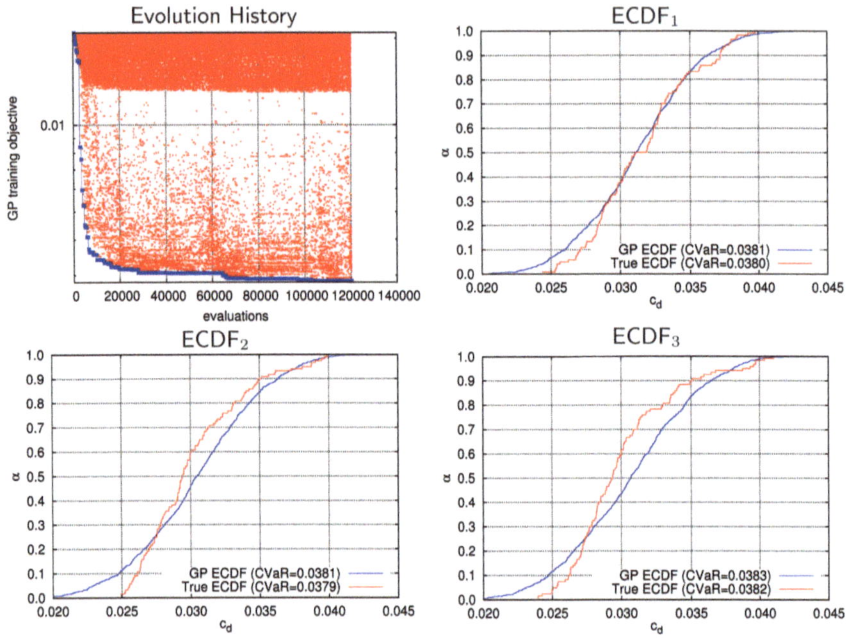

Fig. 7 STEP 4—two points from MC ECDFs test case—evolution history and approximation results

4.4 *STEP 4—Two Points from MC ECDFs*

The last example presented represents a compromise solution. We have selected only two points from the MCs, and the results reported in Table 4 and Fig. 7 show that, this time, the order relationship is maintained and that, presumably, this option represents a good compromise between precision and computational efficiency.

5 Conclusions

In this work, we have shown how two approximation techniques, one intrusive, based on the computation of the gradient through the adjoint field, and a non-intrusive one, based on Gaussian Processes, can be used together to increase the flexibility and efficiency of robust optimization based on the calculation of risk functions using approximated empirical cumulative distributions. In the present work, the newly conceived hybrid approximation technique has been introduced and validated on robust optimization results available in the literature. In the subsequent works, we will apply the method thus obtained to new robust optimization problems in the field of aerodynamic shape design characterized by quantities of interest that are particularly demanding from the computational point of view.

References

1. Dekking, F.M., Kraaikamp, C., Lopuhaä, H.P., Meester, L.E.: A Modern Introduction to Probability and Statistics: Understanding Why and How. Springer (2005). https://doi.org/10.1007/1-84628-168-7
2. Economon, T.D., Palacios, F., Copeland, S.R., Lukaczyk, T.W., Alonso, J.J.: SU2: An open-source suite for multiphysics simulation and design. AIAA J. **54**(3), 828–846 (2016). https://doi.org/10.2514/1.J053813
3. Geuzaine, C., Remacle, J.F.: Gmsh: a 3-D finite element mesh generator with built-in pre- and post-processing facilities. Int. J. Numer. Methods Eng. **79**(11), 1309–1331 (2009). https://doi.org/10.1002/nme.2579
4. Goldberg, D.E.: Genetic Algorithms in Search Optimization and Machine Learning, 1st edn. Addison-Wesley Longman Publishing Co., Inc, USA (1989)
5. Hansen, N.: The CMA evolution strategy: a comparing review. In: Lozano, J.A., Larrañaga, P., Inza, I., Bengoetxea, E. (eds.) Towards a New Evolutionary Computation: Advances in the Estimation of Distribution Algorithms, pp. 75–102. Springer, Berlin, Heidelberg (2006). https://doi.org/10.1007/3-540-32494-1_4
6. Holton, G.A.: Value-at-Risk: Theory and Practice, 2nd edn. Published online at https://www.value-at-risk.net (2013)
7. Morales, E., Bornaccioni, A., Quagliarella, D., Tognaccini, R.: Gradient based empirical cumulative distribution function approximation for robust aerodynamic design. Aerosp. Sci. Technol. **112**, 106630 (2021). https://doi.org/10.1016/j.ast.2021.106630, https://www.sciencedirect.com/science/article/pii/S1270963821001401
8. Morales, E., Quagliarella, D., Tognaccini, R.: Gaussian processes for CVaR approximation in robust aerodynamic shape design. In: Vasile, M., Quagliarella, D. (eds.) Advances in Uncertainty Quantification and Optimization Under Uncertainty with Aerospace Applications, pp. 327–346. Springer International Publishing, Cham (2021)
9. Morales Tirado, E., Quagliarella, D.: Risk Measures in the Context of Robust and Reliability Based Optimization, pp. 411–427. Springer International Publishing, Cham (2021). https://doi.org/10.1007/978-3-030-60166-9_13
10. Quagliarella, D.: Value-at-risk and conditional value-at-risk in optimization under uncertainty. In: Hirsch, C., Wunsch, D., Szumbarski, J., Łaniewski-Wołłk, L., Pons-Prats, J. (eds.) Uncertainty Management for Robust Industrial Design in Aeronautics. Springer, Berlin (2019)
11. Quagliarella, D., Diez, M.: An open-source aerodynamic framework for benchmarking multi-fidelity methods. In: AIAA Aviation 2020 Forum. American Institute of Aeronautics and Astronautics (2020)

12. Quagliarella, D., Iuliano, E.: Robust design of a supersonic natural laminar flow wing-body. IEEE Comput. Intell. Mag. **12**(4), 14–27 (2017)
13. Quagliarella, D., Vicini, A.: A genetic algorithm with adaptable parameters. In: 1999 IEEE International Conference on Systems, Man, and Cybernetics. Institute of Electrical and Electronic Engineers (IEEE), Tokyo, Japan (1999)
14. Rockafellar, R.T., Uryasev, S.: Optimization of conditional value-at-risk. J. Risk **2**, 21–41 (2000)
15. Rockafellar, R.T., Uryasev, S.: Conditional value-at-risk for general loss distributions. J. Bank. Financ. **26**, 1443–1471 (2002)
16. Wiener, N.: The homogeneous chaos. Am. J. Math. **60**(4), 897–936 (1938). https://doi.org/10.2307/2371268
17. Wikipedia Contributors: Empirical distribution function. Wikipedia. https://en.wikipedia.org/wiki/Empirical_distribution_function (2025)
18. Witteveen, J., Doostan, A., Chantrasmi, T., Pecnik, R., Iaccarino, G.: Comparison of stochastic collocation methods for uncertainty quantification of the transonic RAE 2822 airfoil. In: Proceedings of Workshop on Quantification of CFD Uncertainties (2009)

Induced Drag and Vorticity in Viscous and Inviscid Flows

Mauro Minervino and Renato Tognaccini

Abstract Aerodynamic drag prediction and decomposition into physical constituents is a fundamental research topic, and a clear identification of lift-induced drag in viscous and compressible flows is still an open issue. The recent vortical force theory by Wu et al., based on the Lamb vector field integration, allowed for new induced drag definitions to be proposed for viscous and compressible flows, although inconsistencies were reported in their application, with ambiguities on the physical or non-physical role of a non-zero induced drag found in two-dimensional flows. Domain variations of the drag breakdown in induced and parasite contributions also raised questions on the objective or quasi-objective nature of the two drag concepts. In this paper, classical far-field formulas, thermodynamic-based and Lamb-vector-based drag decompositions are analyzed for both inviscid and viscous flows, at low and high Reynolds number, to clearly identify the conditions for induced drag definitions to provide consistent estimates of the drag due to reversible flow phenomena. A spurious induced drag contribution is otherwise computed, resulting in negative induced drag values in two-dimensional flows. This will be shown to be related to the axial velocity perturbation, while thermodynamic methods will confirm the whole drag being of irreversible nature in all analyzed two-dimensional cases, leaving no space to any non-zero reversible drag.

Keywords Induced drag · Vorticity · Lamb vector · Drag decomposition

M. Minervino (✉)
CIRA S.C.p.A. the Italian Aerospace Research Centre, Capua (CE), Italy
e-mail: m.minervino@cira.it
URL: https://www.cira.it/en/

M. Minervino · R. Tognaccini
Industrial Engineering Department, University of Naples Federico II, Naples (NA), Italy
e-mail: renato.tognaccini@unina.it
URL: http://www.dii.unina.it

M. Minervino
e-mail: mauro.minervino@unina.it

M. E. Biancolini et al. (eds.), *Digital Twins for Aerospace*, Springer Aerospace Technology, https://doi.org/10.1007/978-3-032-05022-9_4

1 Introduction

The aerodynamic force which opposes to the motion of a vehicle immersed and advancing in the surrounding fluid (*drag* force) can be ascribed to different physical phenomena and flow features (trailing vortices, viscous wakes and shock waves). An accurate identification of the different drag contributions is of fundamental relevance, for both theoretical and practical reasons. In fact, drag reduction on modern aircrafts (nowadays pushed by highly demanding requirements on the environmental impact of the next-generation fleet) is typically pursued through numerical shape optimization procedures. These, which can exploit information on total drag breakdown to define wiser objective functions [2] and/or to evaluate trade-offs in multi-objective problems [57], improving the design process effectiveness and computational efficiency. On the other hand, advances on aerodynamic force prediction and decomposition theories also allow for a deeper understanding on the genesis of different force contributions, for example *lift-induced* drag, initially identified by the pioneering work of Prandtl at the beginning of the 20th century [41] and related to trailing vorticity (indeed, it is also referred to as *vortex-drag*) [42]. It represents the axial (drag) force associated to the transverse (*lift*) component, generating trailing vortices in three-dimensional flows. The famous Prandtl formula:

$$C_{D,i} = \frac{C_L^2}{\pi \text{Æ} e} \tag{1}$$

can be used to compute the induced drag coefficient ($C_{D,i}$) from the knowledge of lift coefficient (C_L) of lightly-loaded wings with high aspect ratio (Æ) in incompressible flows. However, the Oswald's factor (e) is usually not known a priori unless for very particular cases, and the formula returns zero lift-induced drag at zero global lift coefficient, which is not true in the case of a locally non-zero spanwise lift distribution, with associated trailing vorticity generating drag. For the above reasons, and generally in viscous and compressible flows, other methods are therefore necessary to estimate the lift-induced drag: *far-field* methods.

The standard approach to aerodynamic force ($\boldsymbol{F}$) calculation in Computational Fluid Dynamics (CFD), based on fluid stresses integration on the body surface (∂B) exposed to fluid interaction (*near-field* method), only allows for a mechanical distinction between friction and pressure force contributions:

$$\boldsymbol{F}_{nf} = \int_{\partial B} p\boldsymbol{n}\,\mathrm{d}S - \int_{\partial B} \underline{\underline{\tau}}_v \cdot \boldsymbol{n}\,\mathrm{d}S \tag{2}$$

where p is the static pressure, $\underline{\underline{\tau}}_v$ if the viscous stress tensor and $\boldsymbol{n}$ is a unit normal vector on ∂B, pointing outwards of the adjacent fluidic region. While friction stresses are purely related to viscous effects, wall pressure contributes to both lift-induced drag and *parasite* drag (sum of viscous and wave drag).

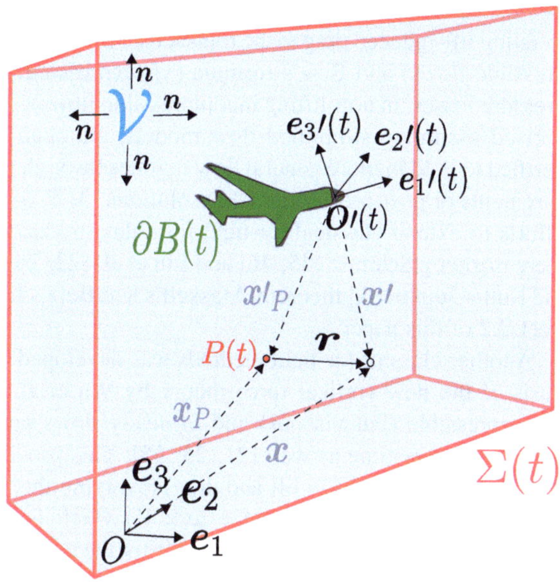

Fig. 1 Generic control volume for far-field force calculation

The phenomenological drag breakdown into viscous, wave and induced components is instead addressed by alternative formulations (far-field methods) based on the integration of flow quantities in a control volume ($\mathcal{V}$) surrounding the body (Fig. 1) and on its external boundary (Σ) [48]:

$$\begin{aligned} \boldsymbol{F}_{ff} = &- \int_{\mathcal{V}} \frac{\partial (\rho \boldsymbol{V})}{\partial t} \mathrm{d}\mathcal{V} - \int_{\partial B} \rho \boldsymbol{V} \boldsymbol{V} \cdot \boldsymbol{n} \, \mathrm{d}S \\ &- \int_{\Sigma} \left(p\boldsymbol{n} - \underline{\underline{\tau}}_v \cdot \boldsymbol{n} + \rho \boldsymbol{V} \boldsymbol{V} \cdot \boldsymbol{n} \right) \mathrm{d}S + \int_{\mathcal{V}} \rho \boldsymbol{f}_m \mathrm{d}\mathcal{V} \end{aligned} \tag{3}$$

where $\boldsymbol{V}$ and ρ are respectively the fluid velocity and the mass density, t is the physical time, and $\boldsymbol{f}_m$ is a body force density per unit mass. An integral balance of linear momentum in $\mathcal{V}$ guarantees the equivalence of near-field and far-field force methods, i.e. $\boldsymbol{F}_{nf} = \boldsymbol{F}_{ff}$. Instead of just using field data at the body skin, far-field approaches directly involve field data from the entire control domain, through volume and surface integrals. In CFD applications, accuracy of far-field force prediction thus requires the flow-field to be fully converged, reducing momentum imbalance to an adequate level. Far-field methods also offer the possibility to visualize the source of the aerodynamic force in the flow domain and to relate it to local flow structures [56] when surface integrals can be converted into volume integrals by applying Gauss's theorem and/or Derivative Moment Transformations (DMTs [36, 55]).

Classical steady far-field methods are represented, for example, by the well-known Kutta–Joukowski theorem [16, 20] (linking lift to velocity circulation, in two-dimensional inviscid and incompressible flow regime), Maskell's formula [28] (relating lift-induced drag to the transverse velocity field on a Trefftz plane, in incompressible flows) and Betz's formula [1] (expressing parasite drag in terms of total pressure losses, in non-lifting incompressible flows). These methods were originally derived assuming simplified flow models, although their validity has been often verified to hold in more general flow regimes through wind-tunnel wake survey measurements or post-processing CFD solutions [3, 5, 6, 13, 14, 19, 51, 52]. Research efforts to extend classical far-field formulas to viscous and/or compressible flows were made by Schmitz [45, 46] and Liu et al. [22, 24]. Generalized formulations of the Kutta–Joukowski theorem, Maskell's and Betz's formulas will also be derived in Sect. 2.2 of this paper.

Another class of far-field methods was developed in the last fifteen years on the basis of the new vortical force theory by Wu et al. [54], initially formulated for incompressible (but viscous) and unsteady flows and later on generalized to the compressible regime as well [21, 23, 55]. This provided exact formulations of the aerodynamic force (both lift and drag) involving the Lamb vector ($\boldsymbol{\ell}$, cross product of flow vorticity and velocity: $\boldsymbol{\ell} = \boldsymbol{\omega} \times \boldsymbol{V}$) which, for steady high-Reynolds number and incompressible flows, is responsible for the generation of the whole aerodynamic force. The theory was further extended to turbulent flows governed by the Reynolds-Averaged Navier-Stokes (RANS) equations by Marongiu and Tognaccini [25] and Mele and Tognaccini [31]. Moreover, a force formula decomposition separating lift-induced and parasite contribution in the total drag (for generally viscous and compressible steady flows) was also initially proposed by Marongiu et al. [27] and Mele et al. [29, 30], then very recently updated by Fournis [11], who also identified links between these Lamb-vector-based expressions of lift-induced/parasite drag and generalized formulations of the Kutta–Joukowski theorem, Maskell's and Betz's formulas as derived and presented in Sect. 2.2.

However, different authors reported inconsistencies using the proposed definitions of lift-induced drag for steady viscous and compressible flows, and negative induced drag values were predicted post-processing two-dimensional flow-fields at low Reynolds numbers [17] or in transonic regime [33], raising doubts on the physical or non-physical role of this non-zero induced drag contribution. Indeed the available far-field definitions of lift-induced drag (e.g. *Maskell*'s formula [28]) do not mathematically ensure zero drag in 2D flows. Moreover, in two-dimensional but *unsteady* flows, the velocity induced by vorticity may give a vortex force component in the free-stream direction, an aspect quite interestingly observed by Saffman [44] (p. 113) when analyzing 2D accelerated inviscid flows. A classical example is the generation of a thrust force on flapping wings and its relation to the Lamb vector fields [37]. Saffman therefore interpreted the birth of induced drag in 3D by analysing a 2D problem, observing that induced drag also exists in 2D (yet unsteady) flows. Nevertheless, in two-dimensional steady conditions, where no trailing vorticity is generated, one usually expects zero induced drag, so the question on whether or not the computed negative values are legitimated is worth a clarification.

On the other hand, prior to Wu's general vortical force theory, thermodynamic-based far-field methods [7, 40] had already been developed at the beginning of the

21th century, moving from Oswatitsch's *entropy drag* concept [39]. These methods, nowadays consolidated and widely used also in the Aviation Industry, only predict drag (lift is usually not addressed) and, in particular, that part of the total drag which is due to flow phenomena increasing fluid entropy (*irreversible* drag), such as viscosity or supersonic compression. Drag due to other *reversible* phenomena (e.g. lift-induced drag) is usually obtained by difference to the total drag, although these methods can also be formulated the other way around, i.e. to directly estimate the reversible drag and indirectly obtaining (by difference) the irreversible part [12, 49, 50].

In this paper, the relationship between Lamb-vector-based lift-induced and parasite drag definitions (directly linked to generalized classical formulas [11]) with reversible and irreversible drag contributions will be theoretically discussed and numerical evidence supporting theoretical argumentation will be provided post-processing two-dimensional CFD solutions around a NACA0012 airfoil in nearly incompressible, subsonic and transonic regime, considering both inviscid and viscous flows, at high or low Reynolds number. Consistency of lift-induced drag formulas with the expected zero reversible drag (predicted by thermodynamic methods) will be shown to hold on very large control volumes only. A spurious induced drag contribution, related to the axial velocity perturbation, is otherwise computed, resulting in negative induced drag values in two-dimensional flows. An application to a three-dimensional elliptic wing in transonic viscous and turbulent flow will also be illustrated, comparing lift-induced drag prediction by Lamb-vector-based methods with results of the celebrated Prandtl formula.

2 Drag Decomposition

We start clarifying the employed definitions for the different forms of aerodynamic drag, identified on a physical and phenomenological basis. Then the *operative* definitions (i.e. the available methods) used herein to predict the different drag contributions will be illustrated and described.

2.1 Drag Definitions

Two different decompositions of the total aerodynamic drag (D) are obtained from the following definitions, summarized by the relation:

$$D = D_{rev} + D_{irr} = D_i + D_p \tag{4}$$

where D_{rev}, D_{irr}, D_i and D_p are defined as:

Reversible Drag (D_{rev}): is the aerodynamic drag obtained in isentropic flow conditions.

It is therefore an *objective* quantity, in the sense that (for example, in steady conditions) its value only depends on the shape and attitude of the body with regards to the fluid it is immersed in, and on flow Mach number (it's worth observing that, for Newtonian fluids with negligible dilatation viscosity coefficient, compressing processes are isentropic as long as the flow regime remains subsonic). However, when a far-field method is used to (directly or indirectly) compute it from a numerical flow solution, a unique reversible drag value cannot be obtained and results depend on the downstream extension of the control volume used by the method. As a matter of fact, in CFD computations another (*spurious*) form of drag is introduced (caused by numerical errors and artificial dissipation of advective fluxes schemes) and the reversible drag content is gradually converted into spurious drag (detected through entropy raising), keeping a constant sum of the two. Indeed, for numerical applications, a spurious drag contribution (D_{sp}) adds to the right-hand-side decomposition in Eq. (4), while on the left-hand-side it is included in D_{irr} as it contributes to entropy production.

Irreversible Drag (D_{irr}): is the drag due to flow phenomena related to entropy production.

In non-isentropic flow conditions, it therefore represents the additional contribution to total drag, adding-on to the reversible drag. Once again, in non-isentropic flow conditions (now even in absence of numerical effects) reversible and irreversible drag are *non-objective* quantities, as the physical dissipation also converts reversible into irreversible drag, with a continuous entropy raise as the flow progresses downstream of the body. Therefore, far-field methods aiming at separating reversible from irreversible drag in a viscous flowfield are expected to predict different breakdowns as the control volume is progressively extended downstream of the body, and usually the decomposition is performed immediately before a significant conversion from reversible into irreversible drag is established.

Lift-Induced Drag (D_i): first identified by Prandtl for inviscid and incompressible flows [41], is the drag (axial aerodynamic force component) directly related to lift (transverse, or *deviating*, component).

Prandtl introduced the concept of *effectively-inviscid* flow ($\mu \to 0$, but $\mu \neq 0$) [42], to justify an aerodynamic force generation not possible for truly inviscid fluids ($\mu = 0$, *d'Alambert paradox*). This fluid model is characterized by an infinitely thin bounded and free vortical sheet which is necessary to guarantee Kutta's condition, but which induces vertical velocities responsible of a downstream axial force on upwards lifting surfaces [41]. Although Prandtl's theory was formulated for incompressible flows and for lightly-loaded wings of high aspect ratio, the concepts at the basis of lift-induced drag generation remain valid for general isentropic flows, with trailing vortices identified as the flow feature this form of drag is related to.

Consistency with Reversible Drag: As no aerodynamic force is generated for truly inviscid fluids according to the d'Alambert's paradox, we define reversible drag only for effectively-inviscid flows and, therefore, only in presence of circulation. The latter in turns implies the presence of a deviating force, then reversible drag can be identified with the drag due to lift (lift-induced drag).

Parasite Drag (D_p): is the drag not directly related to lift, then it represents a drag force with no lift counterpart.

It accounts for losses due to shock waves, boundary layers and viscous wakes. However it's worth recalling that, without viscosity, no aerodynamic force could be generated, therefore parasite drag is indirectly related to lift as well, and inviscid vortex sheets are models for boundary layers and viscous wakes in the vanishing viscosity limit.

Consistency with Irreversible Drag: Based on Eq. (4) and because of consistency between D_{rev} and D_i, this follows as a consequence.

2.2 Far-Field Force Formulas and Operative Drag Definitions

We present three different methods and formulas to compute the aerodynamic force with a far-field approach, together with proposed decompositions of far-field formulas to calculate the different forms of drag previously defined.

The starting far-field force equation (Eq. (3)) will be here considered valid for both laminar and turbulent flow regimes. In the last case, mass density (ρ) and fluid particle velocity ($\boldsymbol{V}$) are intended as their corresponding *ensemble* and *Favre* averages respectively, and a turbulent viscous stress tensor is adopted, including Reynolds stresses ($\underline{\underline{\tau}}_{vt} = \underline{\underline{\tau}}_{v} + \underline{\underline{\tau}}_{R}$). Except for classical methods, no derivation will be here included for the final reported formulas, details being available from the cited literature references.

Thermodynamic Methods Nowadays representing a consolidated approach for drag prediction and decomposition, these methods are usually applied to compute the irreversible drag alone, then reversible drag is indirectly computed by difference to the total drag.

They rely on the possibility to express, for a perfect gas, the velocity magnitude in terms of variations (with regards to asymptotic free-stream conditions) of specific total enthalpy (ΔH), entropy (Δs) and static pressure (Δp) [53] as:

$$\begin{aligned}\frac{V}{V_\infty} &= \sqrt{1 + 2\left(\frac{\Delta H}{V_\infty^2}\right) - \frac{2}{(\gamma-1)M_\infty^2}\left[\left(\frac{\Delta p}{p_\infty}+1\right)^{\frac{\gamma-1}{\gamma}} e^{\frac{\Delta s}{R}\frac{\gamma-1}{\gamma}} - 1\right]} \\ &= f\left(\Delta p/p_\infty, \Delta s/R, \Delta H/V_\infty^2\right)\end{aligned} \quad (5)$$

where M_∞ is the free-stream Mach number, R is the gas constant for the air mixture and γ is the ratio of specific heats.

Focusing on steady flows without body forces (in which case, only the third integral in Eq. (3) is non-zero, for impermeable bodies), Eq. (5) can be used in the axial projection of Eq. (3) to obtain a thermodynamic expression of total drag in terms of Δp, Δs and ΔH (neglecting viscous stresses on Σ, supposing it lays outside of the boundary layer and orthogonal to the free-stream velocity in the body wake).

In this expression, total entropy variations are due to dissipative effects, as well as total enthalpy variations in the case of steady flows and un-powered configurations. Therefore, assuming no impact of dissipative effects on static pressure variations, the irreversible drag contribution is defined as:

$$D_{irr} \stackrel{\text{def}}{=} \boldsymbol{F}_{ff}^{\Delta p=0} \cdot \boldsymbol{e}_x = -V_\infty \int_{\mathcal{V}} \nabla \cdot \left[\rho \hat{v}_x f\left(\frac{\Delta s}{R}, \frac{\Delta H}{V_\infty^2}\right) \boldsymbol{V} \right] \mathrm{d}\mathcal{V} \tag{6}$$

where $\boldsymbol{e}_x$ is a unit vector aligned with $\boldsymbol{V}_\infty$ and the divergence form has been used applying Gauss' theorem. The local axial velocity has been expressed as $\boldsymbol{V} \cdot \boldsymbol{e}_x = V\hat{\boldsymbol{v}} \cdot \boldsymbol{e}_x = V\hat{v}_x$ where $\hat{v}_x$ is related to the angle between local and free-stream velocity vectors (non-negligible in the wake downstream of lifting bodies), which is assumed unitary in the method of Destarac and van der Vooren [7]. The method of Paparone and Tognaccini [40], instead, neglects total enthalpy variations (for a unitary Prandtl number gas) and approximates entropy variation effects using a Taylor series expansion halted at second order terms. In this paper, the method of Destarac and van der Vooren has been applied to isolate and compute the irreversible drag in generally viscous and compressible flow-fields from numerical solutions, and calculated drag values will be denoted as D_{irr}^{DV}.

Finally, it's worth noting that the divergence form of Eq. (6) allows identifying the local generation of irreversible drag in the flow-field (examples will be provided in Sect. 4.4), with possibility to isolate at least part of the so-called *spurious drag* in high-Reynolds number flows, considering any drag generation identified in the irrotational regions of the flow-field [40].

Classical Methods Starting from Eq. (3), taking into account that $\int_\Sigma \boldsymbol{n}\mathrm{dS} = \boldsymbol{0}$ with Σ closed surface, and that $\int_{\mathcal{V}} \frac{\partial \rho}{\partial t}\mathrm{d}\mathcal{V} + \int_{\Sigma \cup \partial B} \rho \boldsymbol{V} \cdot \boldsymbol{n}\mathrm{dS} = 0$ for the continuity equation, it is possible to write:

$$\begin{aligned} \boldsymbol{F}_{ff} = &- \int_{\mathcal{V}} \frac{\partial (\rho \delta \boldsymbol{V})}{\partial t} \mathrm{d}\mathcal{V} - \int_{\partial B} \rho \delta \boldsymbol{V} \boldsymbol{V} \cdot \boldsymbol{n} \, \mathrm{d}S \\ &- \int_{\Sigma} \left[(p - p_\infty)\boldsymbol{n} - \underline{\underline{\tau}}_{vt} \cdot \boldsymbol{n} + \rho \delta \boldsymbol{V} \boldsymbol{V} \cdot \boldsymbol{n} \right] \mathrm{dS} + \int_{\mathcal{V}} \rho \boldsymbol{f}_m \mathrm{d}\mathcal{V} \end{aligned}$$

where we introduced the velocity perturbation $\delta \boldsymbol{V} = \boldsymbol{V} - \boldsymbol{V}_\infty$ (with Cartesian components (u, v, w) along coordinate axes).

The derivation will now proceed assuming steady flows and no body forces. In these hypotheses, volumetric force densities are zero, and convective fluxes at the body skin are also zero for impermeable walls (this is generally not true, even if assuming impermeable bodies, for unsteady flows in the case of moving walls). The total force is therefore expressed in terms of surface integrals only, over the external control volume boundary.

Adding and subtracting $\int_\Sigma \rho(\delta \boldsymbol{V})\boldsymbol{V}_\infty \cdot \boldsymbol{n}\mathrm{dS}$, we obtain:

$$\boldsymbol{F}_{ff} = \int_\Sigma -\left[(p - p_\infty)\boldsymbol{n} - \underline{\underline{\tau}}_{vt} \cdot \boldsymbol{n} + \rho\delta\boldsymbol{V}\delta\boldsymbol{V} \cdot \boldsymbol{n} + \rho(\delta\boldsymbol{V})\boldsymbol{V}_\infty \cdot \boldsymbol{n}\right] \mathrm{dS}$$

Then, considering the double cross-product rule:

$$\boldsymbol{V}_\infty \times \boldsymbol{n} \times \rho\delta\boldsymbol{V} = (\boldsymbol{V}_\infty \cdot \rho\delta\boldsymbol{V})\,\boldsymbol{n} - (\boldsymbol{V}_\infty \cdot \boldsymbol{n})\,\rho\delta\boldsymbol{V}$$

the force formula can be rewritten as:

$$\boldsymbol{F}_{ff} = \int_\Sigma -\left[(p - p_\infty)\boldsymbol{n} - \underline{\underline{\tau}}_{vt} \cdot \boldsymbol{n} + \rho\delta\boldsymbol{V}\delta\boldsymbol{V} \cdot \boldsymbol{n} + (\boldsymbol{V}_\infty \cdot \rho\delta\boldsymbol{V})\,\boldsymbol{n}\right] \mathrm{dS} - \int_\Sigma (\boldsymbol{V}_\infty \times \boldsymbol{n} \times \rho\delta\boldsymbol{V})\,\mathrm{dS}$$

Finally, considering that:

$$\frac{1}{2}\delta\boldsymbol{V} \cdot \delta\boldsymbol{V} = \frac{1}{2}(V^2 - V_\infty^2) - \boldsymbol{V}_\infty \cdot \delta\boldsymbol{V}$$

after some algebra, the far-field aerodynamic force expression can be written as:

$$\begin{aligned}\boldsymbol{F}_{ff} &= \int_\Sigma \left[(p_{t,\infty} - p_t)\boldsymbol{n} + \frac{1}{2}V_\infty^2(\rho - \rho_\infty)\boldsymbol{n} + \underline{\underline{\tau}}_{vt} \cdot \boldsymbol{n}\right] \mathrm{dS} \\ &+ \int_\Sigma \left[-\rho\delta\boldsymbol{V}\delta\boldsymbol{V} \cdot \boldsymbol{n} + \frac{1}{2}\rho(\delta V^2)\boldsymbol{n} + \boldsymbol{V}_\infty \times \boldsymbol{n} \times \rho\delta\boldsymbol{V}\right] \mathrm{dS} \\ &\stackrel{\text{def}}{=} \boldsymbol{F}_{BETZ} + \boldsymbol{F}_{\delta V}\end{aligned} \tag{7}$$

where p_t is the fluid total pressure, the first and second integrals in Eq. (7) have been denoted with the symbols $\boldsymbol{F}_{BETZ}$ and $\boldsymbol{F}_{\delta V}$ respectively (the nomenclature of the first will be justified in the following), and the last term in $\boldsymbol{F}_{\delta V}$ has no drag component, being orthogonal to $\boldsymbol{V}_\infty$.

In non-dissipative flows without shock waves, entropy production is zero and the aerodynamic force exerted in these conditions on the body, if any (effectively ideal flows), has been appellated to as *reversible force* (i.e. force not related to any entropy production). Its component orthogonal and parallel to $\boldsymbol{V}_\infty$ are respectively lift and lift-induced drag.

If the flow also originates from uniform free-stream conditions (omo-enthalpic and omo-entropic flow) and the regime is incompressible, then total pressure variations (first term in $\boldsymbol{F}_{BETZ}$) are zero, and so are density variations (second term) and viscous fluxes (third term). Then $\boldsymbol{F}_{BETZ} = \boldsymbol{0}$ in Eq. (7). In these conditions, exact lift and lift-induced drag formula can be written. Assuming the x-axis aligned with free-stream velocity and z-axis orthogonal to $\boldsymbol{V}_\infty$ and in the direction of the transverse (lift) force ($\boldsymbol{e}_x$ and $\boldsymbol{e}_z$ denote the corresponding unit vectors), it is:

$$\begin{aligned} \boldsymbol{L} &= \left(\boldsymbol{F}_{ff} \cdot \boldsymbol{e}_z\right) \boldsymbol{e}_z = \left(\boldsymbol{F}_{\delta V} \cdot \boldsymbol{e}_z\right) \boldsymbol{e}_z \\ &= \rho_\infty \boldsymbol{V}_\infty \times \int_\Sigma \boldsymbol{n} \times \delta \boldsymbol{V} \mathrm{dS} + \rho_\infty \int_\Sigma \left[\frac{1}{2}(\delta V^2) n_z - w \delta \boldsymbol{V} \cdot \boldsymbol{n}\right] \mathrm{dS}\, \boldsymbol{e}_z \end{aligned} \tag{8}$$

The first term on the right-hand side of Eq. (8) is a three-dimensional Kutta-Joukowski theorem and the second term is a finite domain viscous correction, identified for the first time by Schmitz [46], who also proved that its contribution is zero in a purely irrotational flow. Then, for the axial component:

$$\begin{aligned} D = D_i &= \boldsymbol{F}_{ff} \cdot \boldsymbol{e}_x = \boldsymbol{F}_{\delta V} \cdot \boldsymbol{e}_x = -\int_\Sigma \rho u \delta \boldsymbol{V} \cdot \boldsymbol{n}\, \mathrm{dS} + \int_\Sigma \frac{1}{2} \rho \delta V^2 n_x \mathrm{dS} \\ &= \rho_\infty \int_\Sigma \left(\frac{1}{2} u^2 n_x - u \delta \boldsymbol{V} \cdot \boldsymbol{n}\right) \mathrm{dS} + \rho_\infty \int_\Sigma \frac{1}{2}(v^2 + w^2) n_x \mathrm{dS} \end{aligned} \tag{9}$$

The right-hand side of Eq. (9) is a widely adopted lift-induced drag expression, written for a general external control volume boundary (Σ). It reduces to the more familiar wake integral $\frac{1}{2} \int_W \rho \left(v^2 + w^2 - u^2\right) \mathrm{d}S\, \boldsymbol{e}_x$ when the lateral portion of Σ is sufficiently far from the body and the wake plane W is orthogonal to the freestream velocity. As discussed in [19], Maskell's formula can be obtained from this equation.

An extension of these definitions to generally viscous and compressible flows, however, requires some assumptions. In particular, supposing that:

- the upstream and lateral portions of Σ are not too close to the body and that the flow is not supersonic (so that velocity perturbations can be assumed zero on the lateral portion of Σ, [8])
- the downstream portion of Σ is orthogonal to the free-stream velocity $\boldsymbol{V}_\infty$

then the first two integrals in $\boldsymbol{F}_{BETZ}$ can be restricted on the wake plane, where $\boldsymbol{n} = \boldsymbol{e}_x$. Moreover, the turbulent viscous stress tensor flux (third term in $\boldsymbol{F}_{BETZ}$) on the wake plane can be neglected and therefore $\boldsymbol{F}_{BETZ}$ has no lift component:

$$\boldsymbol{F}_{ff} = F_{BETZ} \boldsymbol{e}_x + \boldsymbol{F}_{\delta V} \tag{10}$$

It represents therefore a drag force with no lift counterpart, i.e. a parasite drag. Then:

$$\begin{aligned} \boldsymbol{L} &= \left(\boldsymbol{F}_{ff} \cdot \boldsymbol{e}_z\right) \boldsymbol{e}_z = \left(\boldsymbol{F}_{\delta V} \cdot \boldsymbol{e}_z\right) \boldsymbol{e}_z \\ &= \boldsymbol{V}_\infty \times \int_\Sigma \boldsymbol{n} \times \rho \delta \boldsymbol{V} \mathrm{dS} + \int_\Sigma \left[\frac{1}{2}\rho(\delta V^2) n_z - \rho w \delta \boldsymbol{V} \cdot \boldsymbol{n}\right] \mathrm{dS} \boldsymbol{e}_z \end{aligned} \tag{11}$$

where the right-hand side of Eq. (11) is a generalized Kutta-Joukowski theorem (for three-dimensional viscous and compressible flows, on finite domains).

If $\boldsymbol{F}_{\delta V} \cdot \boldsymbol{e}_x$ is still assumed as lift-induced drag definition, i.e.:

$$D_i \stackrel{\text{def}}{=} \boldsymbol{F}_{\delta V} \cdot \boldsymbol{e}_x \tag{12}$$

this leads to:

$$\begin{aligned} D &= \boldsymbol{F}_{ff} \cdot \boldsymbol{e}_x = \left(\boldsymbol{F}_{BETZ} + \boldsymbol{F}_{\delta V}\right) \cdot \boldsymbol{e}_x \\ &= \int_\Sigma \left[(p_{t,\infty} - p_t)\boldsymbol{n} + \frac{1}{2} V_\infty^2 (\rho - \rho_\infty)\boldsymbol{n} + \underline{\underline{\tau}}_{vt} \cdot \boldsymbol{n}\right] \mathrm{dS} \cdot \boldsymbol{e}_x \\ &+ \int_\Sigma \left(\frac{1}{2}\rho u^2 n_x - \rho u \delta \boldsymbol{V} \cdot \boldsymbol{n}\right)\mathrm{dS} + \int_\Sigma \frac{1}{2}\rho(v^2 + w^2) n_x \mathrm{dS} \stackrel{\text{def}}{=} D_p + D_i \end{aligned} \tag{13}$$

where the first integral in Eq. (13) is a compressible Betz's formula (justifying the subscript notation used in $\boldsymbol{F}_{BETZ}$) assumed as parasite drag definition:

$$D_p \stackrel{\text{def}}{=} \boldsymbol{F}_{BETZ} \cdot \boldsymbol{e}_x \tag{14}$$

and the second and third integrals represent an extended Maskell's formula, defining lift-induced drag.

The following definitions will be therefore used for the axial and transverse component vectors of $\boldsymbol{F}_{\delta V}$:

$$\boldsymbol{F}_{MSK} \stackrel{\text{def}}{=} \left(\boldsymbol{F}_{\delta V} \cdot \boldsymbol{e}_x\right) \boldsymbol{e}_x = D_i \boldsymbol{e}_x \tag{15}$$

$$\boldsymbol{F}_{KJ} \stackrel{\text{def}}{=} \left(\boldsymbol{F}_{\delta V} \cdot \boldsymbol{e}_z\right) \boldsymbol{e}_z \tag{16}$$

which allow re-writing the classical far-field steady force formula as:

$$\boldsymbol{F}_{ff} = \boldsymbol{F}_{KJ} + \boldsymbol{F}_{MSK} + \boldsymbol{F}_{BETZ} \tag{17}$$

Consistency of Lift-Induced/Parasite versus Reversible/Irreversible Drag Definitions As highlighted before, while the induced drag definition by Eq. (9) gives the reversible drag in the inviscid and incompressible regime, in generally viscous and compressible flows however, a total drag decomposition in lift-induced drag and parasite contributions as defined by Eq. (13), is not consistent with the total drag breakdown in reversible and irreversible components.

Consider, for example, a non-dissipative subsonic flow regime. In this case, there's no entropy production, then the whole drag is *reversible* drag. Nevertheless, D_p as defined in Eq. (13) is non-zero in compressible flows, and therefore D_i does not coincide with the total reversible drag.

We will analyze in more details the relationship between lift-induced/parasite definitions and reversible/irreversible drag breakdown in the following subsection. At this time, it's worth underlying that lift-induced and parasite drag definitions provided by Eqs. (12) and (14) are subject to the hypotheses discussed above for the validity of Eq. (10). A general choice of the control volume would in fact result in non-zero lift component of $\boldsymbol{F}_{BETZ}$, which would modify the general lift expression in Eq. (11) and would not justify the appellation of *parasite* drag attributed to $\boldsymbol{F}_{BETZ} \cdot \boldsymbol{e}_x$.

Vortical Methods The far-field force formula (Eq. (3)), obtained through an integral balance of linear momentum, can be expressed in terms of flow vorticity ($\boldsymbol{\omega}$) by applying Derivative Moment Transformations (DMTs [36, 55]), which in this case represent the integral counterpart of the differential operation of taking the curl of flow momentum equation [56]. The final steady force formula, involving the Lamb vector ($\boldsymbol{\ell} = \boldsymbol{\omega} \times \boldsymbol{V}$), is reported hereafter (denoted with a superscript "b") as derived by Mele and Tognaccini [31] for compressible and statistically stationary flows governed by the RANS equations:

$$\boldsymbol{F}^b_{ff} = \boldsymbol{F}_{\boldsymbol{\ell}} + \boldsymbol{F}_{\boldsymbol{m}_\rho} + \boldsymbol{F}_S + \boldsymbol{F}_\mu \tag{18}$$

where:

$$\boldsymbol{F}_{\boldsymbol{\ell}} = -\int_{\mathcal{V}} \rho \boldsymbol{\ell} \, \mathrm{d}\mathcal{V} - \int_{\partial B} \frac{\rho V^2}{2} \boldsymbol{n} \, \mathrm{d}S \tag{19}$$

$$\boldsymbol{F}_{\boldsymbol{m}_\rho} = -\int_{\mathcal{V}} \boldsymbol{r} \times \left[\nabla\rho \times \nabla\left(\frac{V^2}{2}\right)\right] \mathrm{d}\mathcal{V} - \int_{\partial B} \boldsymbol{r} \times \left[\boldsymbol{n} \times \left(\frac{V^2}{2} \nabla\rho\right)\right] \mathrm{d}S \tag{20}$$

$$\boldsymbol{F}_S = -\int_{\Sigma} \boldsymbol{r} \times [\boldsymbol{n} \times (\rho\boldsymbol{\ell})] \, \mathrm{dS} \tag{21}$$

$$\boldsymbol{F}_\mu = \int_{\Sigma} \boldsymbol{r} \times \left(\boldsymbol{n} \times \nabla \cdot \underline{\underline{\tau}}_{vt}\right) \mathrm{dS} + \int_{\Sigma} \underline{\underline{\tau}}_{vt} \cdot \boldsymbol{n} \, \mathrm{dS} + \int_{\mathcal{V}} \boldsymbol{r} \times \left(\nabla \times \tilde{\delta}_{\rho V}\right) \mathrm{d}\mathcal{V} \tag{22}$$

$\boldsymbol{F}_{\boldsymbol{\ell}}$ is the *vortex force* (namely, the volume integral of the opposite Lamb vector, times mass density, in $\mathcal{V}$), corrected by a body surface term which is non-zero for inviscid flows [17]. $\boldsymbol{F}_{\boldsymbol{m}_\rho}$ is a compressibility correction, taking into account density variations in the flow-field. $\boldsymbol{F}_S$ represents the contribution of flow vorticity from outside of the selected control volume $\mathcal{V}$ and $\boldsymbol{F}_\mu$ includes terms which are explicitly depending on turbulent fluctuations and flow viscosity (through the laminar viscous stress tensors $\underline{\underline{\tau}}_v$ and Reynolds stresses $\underline{\underline{\tau}}_R$).

The symbol $\boldsymbol{r}$ denotes a scaled position vector, computed from an arbitrary pole (Fig. 1):

$$\boldsymbol{r} = \frac{\boldsymbol{x} - \boldsymbol{x}_{POLE}}{\mathcal{N} - 1} \tag{23}$$

with $\mathcal{N}$ being the space dimension (2 or 3). In the turbulent term:

$$\tilde{\boldsymbol{\delta}}_{\rho V} = \overline{\boldsymbol{V}''\nabla \cdot (\rho \boldsymbol{V}'')} \tag{24}$$

the ensemble average is indicated with $\bar{\ }$ and the *Favre* average with $\tilde{\ }$, while $''$ denotes random properties fluctuations with respect to the *Favre* average.

Differently from the classical far-field force expression (Eq. (7) or Eq. (17)) which, for steady flows, is only based on surface integrals, vortical force formulas as Eq. (18) also include domain integrals in $\mathcal{V}$, a feature which allows linking the aerodynamic force to its local generation in the flow-field (examples will be provided in Sect. 4.4).

In this paper, a modified vortical formula derived by Mele et al. [30] will be used (denoted with a superscript "*c*"), as it provided higher accuracy in the numerical post-processing of transonic flow-fields [33, 34]. In this formula, which is mostly based on surface integrals on Σ (as for classical methods), the vortex force contribution ($\boldsymbol{F}_\ell$) and the compressibility correction term ($\boldsymbol{F}_{m_\rho}$) are combined together in the mathematically equivalent expression [30]:

$$\boldsymbol{F}_{\ell+\rho} = -\int_{\Sigma} \rho \boldsymbol{V}\boldsymbol{V} \cdot \boldsymbol{n}\, \mathrm{dS} + \int_{\Sigma} \rho \frac{V^2}{2} \boldsymbol{n}\, \mathrm{dS} + \int_{\Sigma} \boldsymbol{r} \times \left[\boldsymbol{n} \times \left(\frac{V^2}{2} \nabla \rho\right)\right] \mathrm{dS} \tag{25}$$

leading to the final formula written in compact form:

$$\boldsymbol{F}^c_{ff} = \boldsymbol{F}_{\ell+\rho} + \boldsymbol{F}_S + \boldsymbol{F}_\mu \tag{26}$$

Mele et al., working on a mixed "*b*/*c*" force formula, also proposed a decomposition of the total drag in lift-induced and parasite contributions [30], which was later on updated by Fournis et al. [11] to guarantee drag breakdown independence from the arbitrary pole appearing in Eq. (23). Their proposed definitions were adapted to the "*c*" force formula by Minervino and Tognaccini [33, 34] and are reported hereafter:

$$D^c_i \stackrel{\text{def}}{=} \left(\boldsymbol{F}_{\ell+\rho} - \boldsymbol{F}_{\nabla\rho}\right) \cdot \boldsymbol{e}_x \tag{27}$$

$$D^c_p \stackrel{\text{def}}{=} \left(\boldsymbol{F}_S + \boldsymbol{F}_\mu + \boldsymbol{F}_{\nabla\rho}\right) \cdot \boldsymbol{e}_x \tag{28}$$

where:

$$\boldsymbol{F}_{\nabla\rho} = \frac{V_\infty^2}{2} \int_{\Sigma} (\rho - \rho_\infty)\, \boldsymbol{n} \mathrm{d}S + \int_{\Sigma} \boldsymbol{r} \times \left(\boldsymbol{n} \times \frac{V^2}{2} \nabla \rho\right) \mathrm{d}S \tag{29}$$

is an additional compressibility correction that gradually vanishes as the control volume external surface (Σ) extends away from the body [10].

Links to Classical Formulas: Fournis et al. [11] also identified the links between the vortex-force theory and classical far-field methods (the Kutta-Joukowski theorem, Maskell's formula and Betz's formula) proving the following notable identities:

$$\boldsymbol{F}_{\ell+\rho} - \boldsymbol{F}_{\nabla\rho} = \boldsymbol{F}_{KJ} + \boldsymbol{F}_{MSK} \tag{30}$$

$$\boldsymbol{F}_S + \boldsymbol{F}_\mu + \boldsymbol{F}_{\nabla\rho} = \boldsymbol{F}_{BETZ} \tag{31}$$

Consistency of Lift-Induced/Parasite versus Reversible/Irreversible Drag Definitions: Because of the links with classical methods discussed above, considerations from the previous subsection similarly apply to vortical methods. Nevertheless, the capability of lift-induced and parasite drag definitions by Eqs. (27) and (28) to respectively represent reversible and irreversible drag contributions is further discussed herein, considering two particular conditions:

- **subsonic inviscid flow**: the total drag (D) is in this case due to reversible flow phenomena associated with the generation of trailing vorticity in three-dimensional lifting cases:

  $$D = D_{rev}$$

  However, decomposing total drag in lift-induced and parasite contributions:

  $$D = D_i + D_p$$

  and using definitions from Eqs. (27) and (28), $\boldsymbol{F}_\mu = 0$ (the flow is inviscid) and $\boldsymbol{F}_S = 0$ (the Lamb vector is zero in the free vortex system, where Kelvin's theorem ensures that flow vorticity is aligned with the fluid velocity) but $\boldsymbol{F}_{\nabla\rho}$ is non-zero unless Σ is not sufficiently far from the body skin. Therefore, parasite drag as defined by Eq. (28) is non-zero and lift-induced drag (Eq. (27)) does not coincide with total reversible drag, unless Σ retreats to infinity.
- **no trailing vorticity**: the total drag (D) is in this case due to irreversible flow phenomena associated with the generation of boundary layers, shock waves and their wakes. The irreversible drag can be expressed as the sum of Destarac and van der Vooren (DV) definition [7] (which accounts for entropy and total enthalpy variations of the streamwise velocity field) plus an additional term (ΔD_{irr}) accounting for the irreversible drag not included in DV definition (associated to static pressure variations and transverse flow velocities). The last term reduces to zero as Σ retreats to infinity as, in the absence of trailing vorticity, zero transverse flow disturbances and free-stream static pressure are realized in the body far wake. Then:

  $$D = D_{irr} = D_{irr}^{DV} + \Delta D_{irr}; \qquad \lim_{\Sigma\to\infty} \Delta D_{irr} = 0$$

However, in a very recent work [9] Fournis highlighted how, in transonic inviscid flows, parasite drag defined by Eq. (28) does not coincide with D_{irr}^{DV}, and the difference (ΔD_p) approaches zero only when Σ retreats to infinity:

$$D_p = D_{irr}^{DV} + \Delta D_p; \qquad \lim_{\Sigma \to \infty} \Delta D_p = 0$$

Therefore, decomposing total (irreversible) drag in lift-induced and parasite drag as proposed in Eqs. (27) and (28), and using the above relations:

$$D = D_{irr}^{DV} + \Delta D_{irr} = D_i + D_p = D_i + D_{irr}^{DV} + \Delta D_p$$

leads to non-zero lift-induced drag in this case ($D_i = \Delta D_{irr} - \Delta D_p$), unless Σ is chosen sufficiently far from the body skin.

Moreover, it has to be mentioned that definitions provided in Eqs. (27) and (28) are linked respectively to an extended Maskell's formula ($\boldsymbol{F}_{MSK}$) and to a compressible Betz's formula ($\boldsymbol{F}_{BETZ}$, Sect. 2.2) where the contribution of $\int_\Sigma \rho \left(\frac{u^2}{2} n_x - u(\delta \boldsymbol{V} \cdot \boldsymbol{n}) \right) \mathrm{d}S\boldsymbol{e}_x$ has been included in $\boldsymbol{F}_{MSK}$ instead of being attributed to parasite drag ($\boldsymbol{F}_{BETZ}$), as suggested by Maskell [28]. This contribution (which reduces to the more familiar wake integral $\frac{1}{2} \int_W -\rho u^2 \mathrm{d}S\boldsymbol{e}_x$ when the lateral portion of Σ is sufficiently far from the body and the wake plane W is orthogonal to the freestream velocity) is questioned, and has been analyzed by Spalart [47] and Drela [8]. The former also discussed the possibility of the $-\rho u^2$ term to provide negative lift-induced drag (as allowed by the formula), a feasible circumstance at least in (idealized) two-dimensional flows. The latter, justified its attribution to the induced drag in non-rotational flow regions of the Trefftz plane based on kinetic energy considerations and analyzing incompressible flow regimes. However, in rotational Trefftz plane regions (where vortex wakes are mixed with viscous and shock wakes) and/or in presence of compressibility effects, the role of this term is still ambiguous and so is drag decomposition. This ambiguity can be eliminated if Σ retreats to infinity in the linear far-field, as the abovementioned drag contribution is in this case negligible.

3 Numerical Test-Cases

CFD solutions were obtained for a NACA0012 airfoil with unitary chord length (c) using the *ZEN* flow solver developed at the Italian Aerospace Research Centre (CIRA) [26]. CIRA *ZEN* code is a multi-block structured flow solver, based on a cell-centered formulation with central schemes and *Jameson*-like artificial dissipation model [15].

Weak Euler solutions are computed in the case of effectively inviscid flow simulations ($\mathrm{Re}_\infty \to \infty$), while, in viscous flow simulations at low Reynolds num-

Table 1 Simulated flow conditions, NACA0012 airfoil

	Nearly-incompressible $M_\infty = 0.05$	Subsonic $M_\infty = 0.5$	Transonic $M_\infty = 0.8$
Weak Euler: $\mathrm{Re}_\infty \to \infty$	$\alpha = 0°$	$\alpha = 0°$	$\alpha = 0°$
	$\alpha = 4°$	$\alpha = 4°$	$\alpha = 2°$
RANS: $\mathrm{Re}_\infty = 9 \times 10^6$	$\alpha = 0°$	$\alpha = 0°$	$\alpha = 0°$
	$\alpha = 4°$	$\alpha = 4°$	$\alpha = 3°$
NS: $\mathrm{Re}_\infty = 1 \times 10^4$		$\alpha = 2°$	
		$\alpha = 4°$	

ber ($\mathrm{Re}_\infty = 1 \times 10^4$), Navier-Stokes (NS) equations were solved assuming laminar flow. High Reynolds number analyses ($\mathrm{Re}_\infty = 9 \times 10^6$) used a Reynolds-Averaged Navier-Stokes (RANS) formulation of flow governing equations, with turbulence closure based on $k - \omega$ turbulent/non-turbulent (TNT) model [18]. In this case, fully turbulent flow was assumed and pre-conditioning was used for the nearly incompressible case simulations.

A slightly altered definition of the NACA0012 airfoil providing closed trailing edge is used, according to the procedure illustrated in [43]. The computational mesh is a C-type grid with $828 \times 200 = 165{,}600$ hexahedral cells (428 around the airfoil) generated using NASA *HYPGEN* hyperbolic grid generator [4]. This will be referred to in the following text as "fine mesh". In viscous flow simulations, cells are clustered around the airfoil surface to allow computing numerical solutions down to the wall without using any wall function ($y^+ \approx 1$). The computational domain extends up to 300 chord lengths around the airfoil for subsonic simulations, while for transonic cases the reference fine mesh was modified enlarging the computational domain size by a factor 10. At the scope, additional cell layers were included in the original mesh to cover the region between the initial far-field boundary location up to the extended domain boundary ($\approx 3000c$).

Numerical solutions were computed at two different angles of attack for different flow regimes according to Table 1. Convergence of flow solution to machine accuracy was achieved, although, for viscous simulations at zero angle of attack, computations revealed particularly stiff, requiring a much larger number of solver iterations, compared to other cases, to achieve a similar level of convergence. A grid convergence study is already available in [34].

A steady three-dimensional test-case was also considered, analyzing an elliptic-planform wing with no twist, straight trailing edge and $\mathcal{A}\!\!R = 7$. Wing sections are shaped using a NACA0012 airfoil with closed t.e. and the chord length at mid-span of the wing (c_0) is unitary. Calculations were performed in viscous transonic flow conditions ($M_\infty = 0.75,\ \mathrm{Re}_\infty = 3 \times 10^6$) and at zero side-slip angle ($\beta = 0$).

Being the flow-field symmetrical with regards to the mid-plane at zero side-slip angle, a half of the wing configuration was analyzed. The grid topology is of C-C

Fig. 2 Computational grid used for the 3D elliptic wing simulations: surface mesh visualization at the body wall and on orthogonal grid planes

type (Fig. 2), structured in two symmetrical blocks (upper-side and lower-side) with opposite node coordinates in the direction normal to the wing planform. The far field distance from the body is set at 30 root chord lengths. The volume mesh contains more than 3×10^6 hexahedral cells (384×64 quadrilateral elements over the wing surface, 64 nodes downstream in the wake and 96 interior nodes in the direction

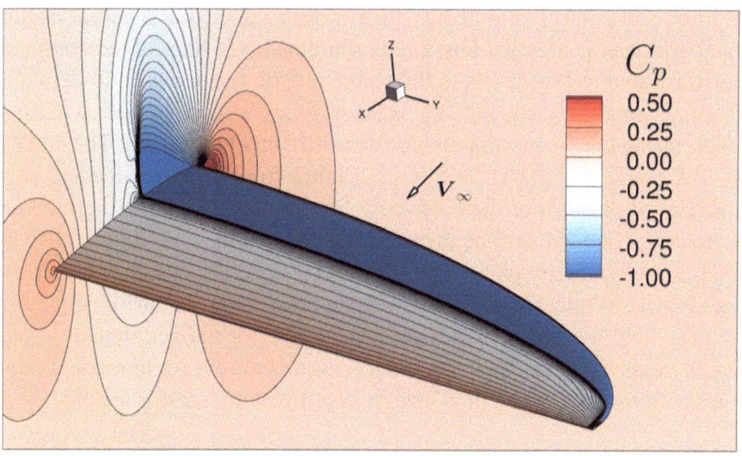

Fig. 3 Contour plot of local pressure coefficient (on the wing surface and on the symmetry plane) for the 3D steady RANS simulation in viscous transonic conditions ($M_\infty = 0.75$; $\mathrm{Re}_\infty = 3 \times 10^6$; $\alpha = 4°$)

normal to the wall/wake). For the same computational mesh, a grid convergence study is already available in [29]. Also in this case, convergence to machine accuracy was achieved (12-orders of magnitude drop, for the average mass-flux residual) and a unitary non-dimensional wall distance was obtained at the first cell layer ($y^+ \approx 1$; Fig. 3).

4 Numerical Results

The aerodynamic force (lift and drag) was computed, in the post-processing phase, using both near-field surface stress integration (Eq. (2)) and the far-field Lamb-vector-based method illustrated in Sect. 2 (Eq. (26)). For the latter, the only term that was not included in computations is the last integral of Eq. (22), which is related to turbulent fluctuations and represents an unclosed term. This was neglected assuming *Morkovin*'s hypothesis stating that compressibility has a weak influence on the Reynolds stress [35]. Total drag was decomposed in lift-induced and parasite contributions, computed using definitions in Eqs. (27) and (28). Then the irreversible drag was also calculated through Eq. (6) using the method of Destarac and van der Vooren [7]. Lift and drag coefficients are referred to the free-stream dynamic pressure and to the unitary chord length ($C_l = L/(0.5\rho_\infty V_\infty^2 c)$ and $C_d = D/(0.5\rho_\infty V_\infty^2 c)$, respectively).

Flow vorticity required in Lamb vector calculation has been computed using a nodes-based Green-Gauss formula, and volume integrals are numerically computed by simply summing cell center values, multiplied by the cell volume, over the selected integration domain. Similarly, surface integrals are obtained computing the state in the cell surface center obtained by centered averaging.

The employed control volume has always been selected as a subset of the computational grid, and is identified by three orthogonal grid lines of the computational mesh as illustrated in Fig. 4, where the characteristic lengths used to define its size are also represented. In particular, X_W measures the chordwise position of the "wake plane" delimiting the integration domain downstream of the airfoil. The downstream portion of Σ (wake plane) has chordwise distance from the airfoil leading edge X_W, while the lateral distance of the C-shaped Σ edge from the airfoil will be denoted as Z_W. Both X_W and Z_W will be expressed in chord units and, for the 3D wing, are measured on the symmetry plane. The DMT pole (Fig. 1) is located at the airfoil leading edge (Fig. 4). While, for given free-stream conditions and airfoil angle of attack (α), the near-field method provides a unique value for the aerodynamic force, the theoretically equivalent far-field calculation might exhibit, for numerical reasons, a dependence on some of the method's degrees of freedom, such as the control volume choice.

In the analyzed cases, the total force revealed being very weakly influenced by the adopted integration domain, being substantially constant at condition that the control volume extends downstream of the airfoil by at least three or four chord lengths, depending on the analyzed case. Figures 5 and 6 illustrate the difference

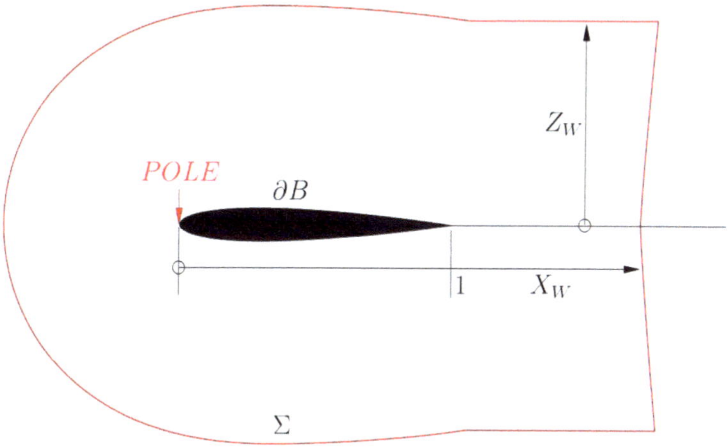

Fig. 4 Control volume shape and characteristic dimensions

between 2D far-field and near-field computed lift and drag coefficients, where the first are calculated for a variable control volume size, which dimension is modified varying both X_W and Z_W simultaneously (i.e. $X_W = Z_W$ and $X_W - 1$ is represented on the abscissa, using a logarithmic scale). Lift coefficients rapidly converge well below one lift count accuracy (± 0.01) for Σ distant more than a half chord length from the airfoil (Fig. 5), while drag coefficients convergence is achieved after three or four chord lengths within two drag counts ($d.c.$) accuracy (± 0.0002), except for low Reynolds number simulations, where the difference stays around 10 $d.c.$ (Fig. 6). This was explained by a laminar flow separation observed at both incidence angles ($\alpha = 2°$ and $\alpha = 4°$) which influences the flow-field convergence and therefore the numerical satisfaction of flow momentum balance.

Large differences are also observed for very small control volumes (in the region close to the wall, where local residuals are larger) as well as for control volumes smaller than approximately one chord length in transonic conditions or in presence of laminar separation, when the shock wave or the separation region (sources of local numerical inaccuracies) are not fully enclosed in the control volume and intercepted by its external surface, polluting results of Σ-integrals.

4.1 Reversible Drag, in 2D

For two-dimensional flows, where no trailing vortices are realized, the expected reversible drag is zero, and total drag is entirely due to irreversible flow phenomena (viscosity or supersonic compression) increasing fluid entropy. Figure 7 illustrates the difference between total far-field drag (computed from the vortical formula in

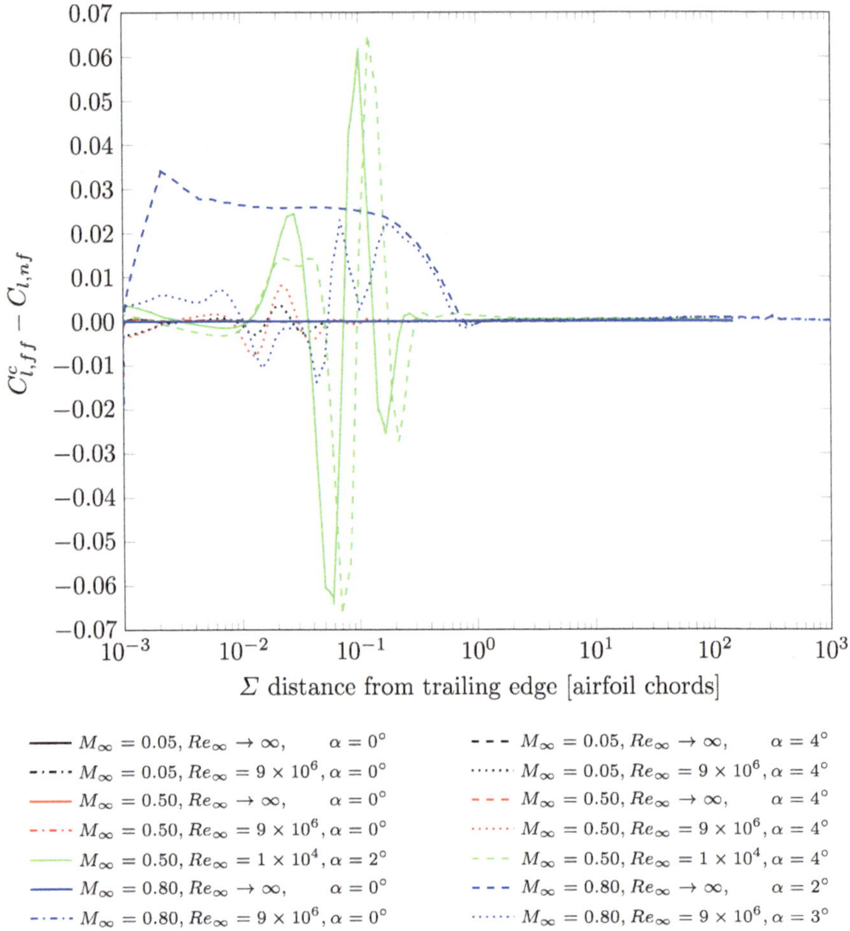

Fig. 5 Far-field versus near-field lift coefficient difference as a function of the integration domain size, for all simulated 2D flow conditions

Eq. (26)) and the irreversible drag (computed using the method of Destarac and van der Vooren [7]): an indirect reversible drag measure.

It is clear how they converge onto each other within one $d.c.$ accuracy in the same range where far-field drag was observed to stabilize (after three or four chord lengths). This numerically confirms the irreversible nature of total drag for the analyzed two-dimensional flows.

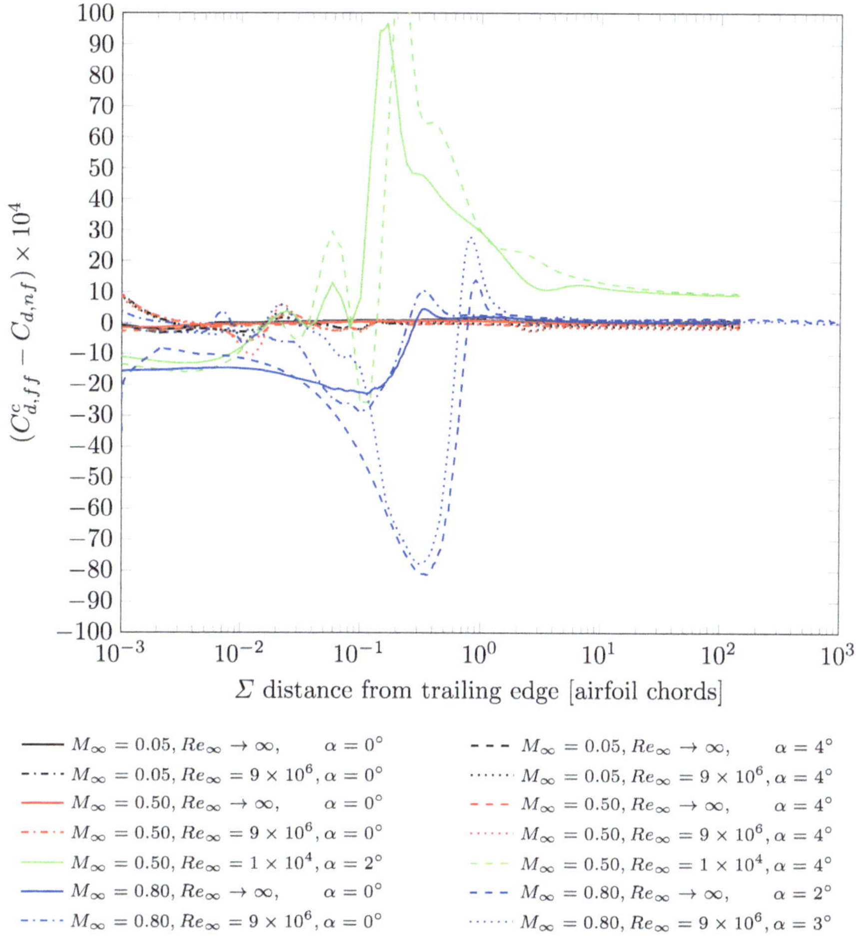

Fig. 6 Far-field versus near-field drag counts difference as a function of the integration domain size, for all simulated 2D flow conditions

4.2 *Lift-Induced Drag, in 2D*

The consistency of total drag decomposition in lift-induced and parasite contributions with the same drag breakdown in reversible and irreversible parts, theoretically discussed in Sect. 2.2, has a numerical evidence in the results illustrated in Fig. 8, where lift-induced drag coefficients (computed using Eq. (27)) are plotted as a function of the integration domain size. It is clearly visible that, in most of simulated cases (black curves), lift-induced drag rapidly approaches zero and is negligible ($-1\,d.c. < C^c_{d,i} < +1\,d.c.$) for a Σ distance greater than 5 airfoil chords. On the other hand, in stiff numerical simulations (blue curves), transonic asymmetric cases (red curves) or low Reynolds number conditions (green curves), the

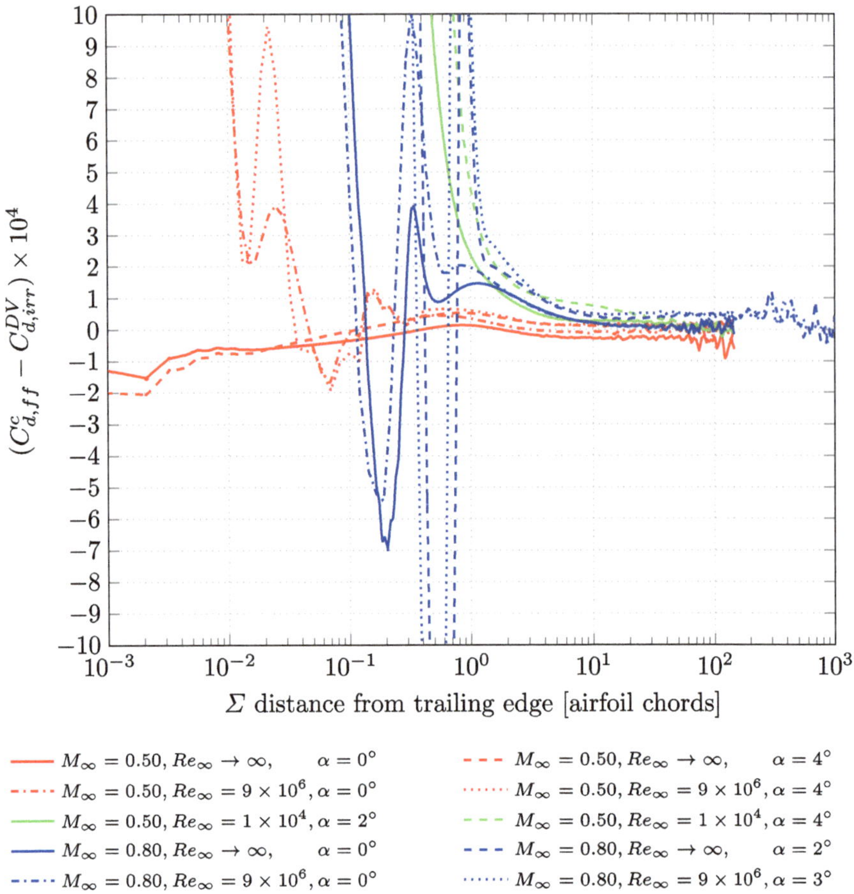

Fig. 7 Far-field versus near-field drag counts difference as a function of the integration domain size, for all simulated 2D flow conditions

convergence of lift-induced drag to the (null) reversible drag is much slower and significant negative values of such a *spurious* lift-induced drag are still obtained for a control volume size in the order of 10 or 100 chords. In particular, for the two transonic cases with non-zero incidence, data included in Fig. 8 are related to an extended computational domain with far-field boundary at $\approx 3000c$ instead of $\approx 300c$, to highlight how zero lift-induced drag is indeed achieved in these cases on very large control volumes (Σ distance $\approx 1000c$). At low Reynolds number conditions, results are in line with what was already observed by Kang et al. [17]. They used formulation "*b*" of lift-induced drag (without the additional compressibility correction $\boldsymbol{F}_{\nabla\rho}$, which is however negligible in this case, although here not illustrated) and a wake plane location between 2 and 3 chord lengths downstream of the airfoil trailing edge, and observed negative induced drag coefficients in the order of

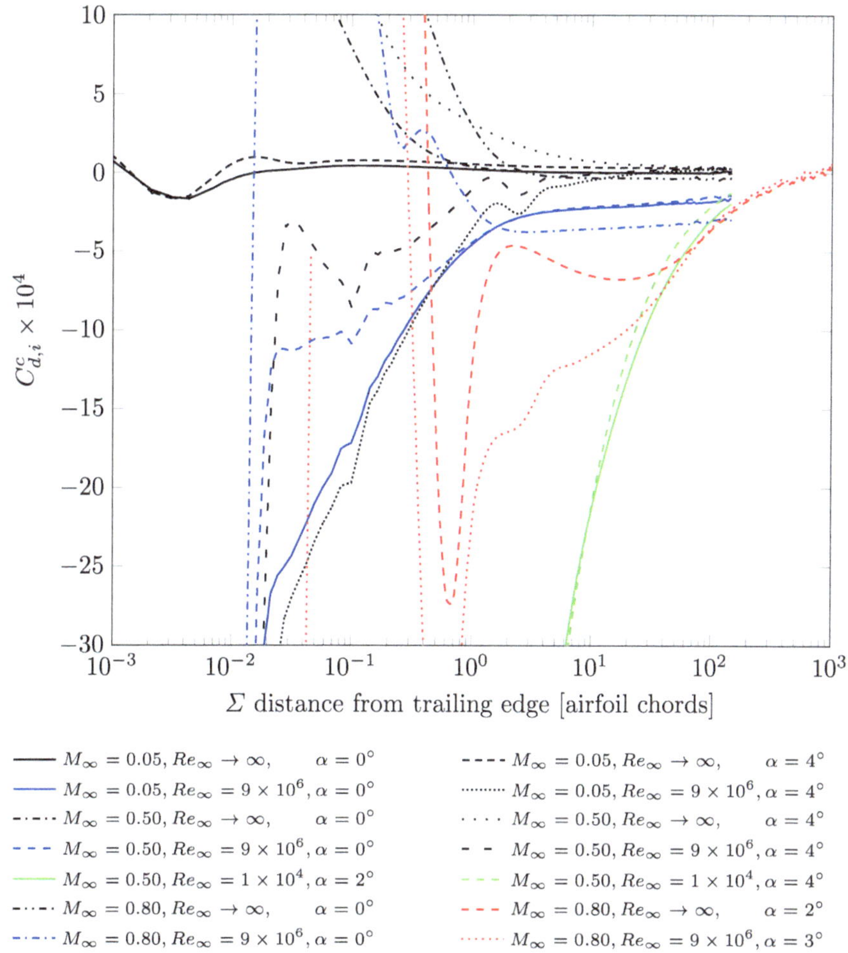

Fig. 8 Far-field versus near-field drag counts difference as a function of the integration domain size, for all simulated 2D flow conditions

approximately $-100\,d.c.$ at $\mathrm{Re} = 1 \times 10^4$ ([17], Fig. 21, p. 4636). This value however provides a rough estimate only, as they simulated subsonic flow conditions at $M = 0.3$ (instead of $M = 0.5$) and $\mathrm{Re} = 1 \times 10^3$, $\mathrm{Re} = 1 \times 10^6$ and $\mathrm{Re} = 1 \times 10^7$ (i.e. value at $\mathrm{Re} = 1 \times 10^4$ is interpolated).

The decomposition in lift-induced and parasite contributions is therefore not independent on the selected control volume in the highlighted cases, even in absence of trailing vorticity (2D flows), and consistency with total drag breakdown in reversible and irreversible parts (which, instead, stabilize quite quickly downstream of the airfoil near wake) is only achieved on very large integration domains.

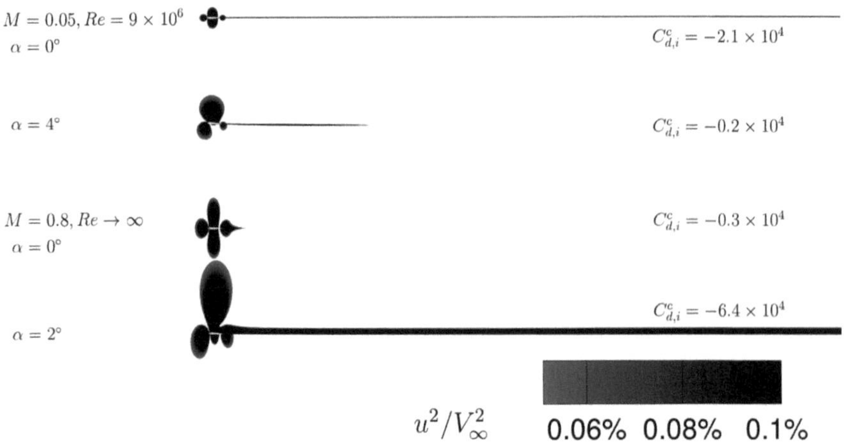

Fig. 9 Contour plot of non-dimensional streamwise velocity perturbation (squared) in nearly incompressible viscous flow solutions and inviscid transonic flow solutions ($C^c_{d,i}$ computed with Σ distance $\approx$ 15 airfoil chords)

Negative Lift-Induced Drag and Axial Velocity Perturbation Figure 9 also illustrates, for some of the analyzed cases, how the different rates of convergence for lift-induced drag observed in Fig. 8 correspond to different decay rates of the streamwise velocity perturbation (u).

4.3 3D Elliptic Wing

In the previous subsections, it was observed how an accurate drag breakdown in lift-induced and parasite contribution requires control volumes extending far away downstream of the body. Unfortunately, in numerical applications and especially in 3D, the integration domain cannot be extended too far downstream of the body, because a spurious conversion of reversible into irreversible drag also occurs as discussed in Sect. 2.1.

In Fig. 10 the drag polar was computed for the elliptic wing case described in Sect. 3 by direct surface stress integration (near-field method), and compared with the same data obtained by the Lamb-vector-based far-field formula (18). In this case, despite the presence of shock waves in the flow-field, the accuracy of Eq. (18) was recovered using the alternative Lamb-vector definition described in [32], instead of using the transonic formula (26) as done for two-dimensional test cases.

In the same figure, the total far-field drag is decomposed isolating the lift-induced contribution. Results provided by the incompressible Prandtl formula are also included in the plot, being the Oswald factor known and unitary (elliptic planform) and considering that the wing has no aerodynamic twist. Computed induced drag values are very close to the reference incompressible Prandtl curve, although smaller by a few drag counts.

The control volume selected for the drag decomposition illustrated in Fig. 10 extends approximately 4.5 root chord lengths downstream of the wing trailing edge. As a matter of fact, the computed induced drag is underestimated as affected by the spurious negative contribution already observed in two-dimensional flows. This effect is stronger in the near wing's wake (where the axial velocity defect is higher) and slowly vanishes downstream, reducing the underestimation. On the other hand, the reversible drag content of the flow-field is also gradually reducing, due to the physical and numerical dissipation, then the computed lift-induced drag has a maximum when this effect overcomes the former (Fig. 11). A typical choice, in Lamb-vector methods, is to assume the peak value as best estimate for the lift-induced drag. The reversible drag contribution was also calculated through the method of Destarac and van der Vooren [7], Eq. (26), by difference to the total near-field drag and plotted in Fig. 11. At the trailing edge of the wing, the computed reversible drag is surprisingly close to the value predicted by Prandt's incompressible formula, then it gradually decreases as the wake plane is moved downstream of the wing, due to the reasons discussed above. The simulation at zero incidence (and zero lift) is instead analyzed in Fig. 12. A similar behaviour is observed for the lift-induced and reversible drag variations with the wake plane position, the main difference consisting in the absence of reversible drag ($C_{D,i} = 0$ is also predicted by Prandtl's formula) and, therefore, in the lack of conversion into irreversible drag. Nevertheless, the computed lift-induced drag is not able to approach the zero value as the spurious negative contribution vanishes very slowly (a logarithmic scale was used in Fig. 8) and in this case the computational domain only extends 30 root chord lengths downstream of the wing.

4.4 Drag Visualizations

It has been mentioned that, using Lamb-vector-based vortical far-field methods, the local generation of aerodynamic force in the flow-field can be identified. At the scope, force formulas based on volume integrals are necessary. Equation (18) introduced in Sect. 2.2 still uses surface integrals in the last two terms ($\boldsymbol{F}_S$ and $\boldsymbol{F}_\mu$). The contribution of the latter is however negligible for high-Reynolds number flows, supposing the Σ surface does not fully lie within the viscous boundary layer [25, 31]. An equivalent formulation for the $\boldsymbol{F}_S$ term, valid for viscous flows and based on domain integrals, has been derived by Ostieri and Tognaccini [38] and later on extended to inviscid flows by Minervino and Tognaccini [33]. Details will not be reported herein, but this equivalent expression, indicated with the symbol $\boldsymbol{F}_{S\mathcal{V}}$, can be used in the total force formula (denoted with a superscript "d") which then writes as:

$$\boldsymbol{F}^d_{ff} = \boldsymbol{F}_\ell + \boldsymbol{F}_{m_\rho} + \boldsymbol{F}_{S\mathcal{V}} + \boldsymbol{F}_\mu \approx \boldsymbol{F}_\ell + \boldsymbol{F}_{m_\rho} + \boldsymbol{F}_{S\mathcal{V}} \tag{32}$$

where the right-hand side only contains volume integrals. On a discrete basis, instead of visualizing the local force density (integrand functions), a contour plot of the elementary contribution to domain integrals of each grid cell of the computational

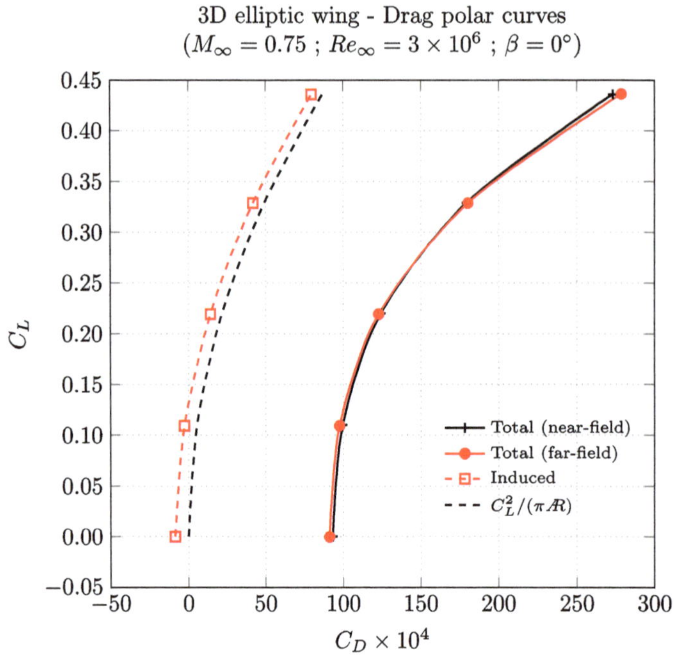

Fig. 10 Drag polar breakdown isolating lift-induced contribution, for the 3D elliptic-planform wing, computed using a Lamb-vector far-field force formula ($X_W/c_0 = 4.5$, $Z_W/c_0 = 10.1$)

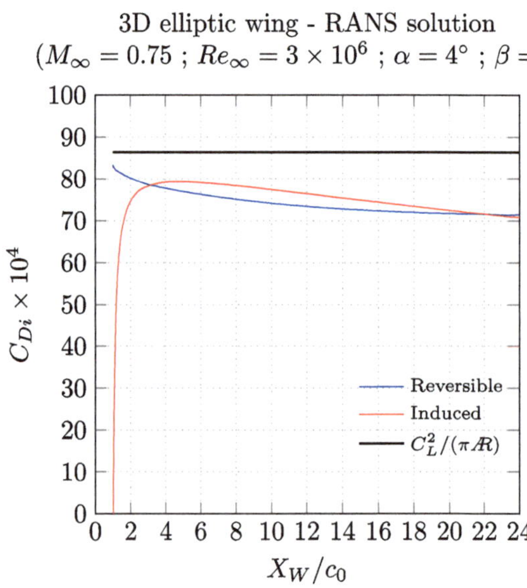

Fig. 11 Lift-induced and reversible drag of the 3D elliptic-planform wing at non-zero incidence (lifting case), compared to Prandtl formula value

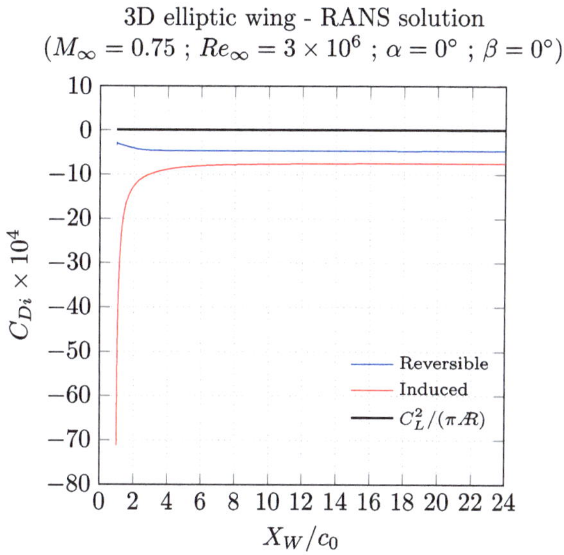

Fig. 12 Lift-induced and reversible drag of the 3D elliptic-planform wing at zero incidence (non-lifting case) compared to Prandtl formula value

domain is illustrated in Fig. 13 for total lift and in Fig. 14 for total drag, in viscous transonic conditions at non-zero incidence. The aerodynamic force is only generated in the rotational flow-field regions (boundary layer, viscous wake, shock wave and shock wake).

Force decomposition formulas, equivalent to Eqs. (27) and (28), have also been derived [29, 34] and can be written as:

$$D_i^d \stackrel{\text{def}}{=} \left(\boldsymbol{F}_{\boldsymbol{\ell}} + \boldsymbol{F}_{\boldsymbol{m}_\rho} - \boldsymbol{F}_{\nabla\rho}\right) \cdot \boldsymbol{e}_x \approx \left(\boldsymbol{F}_{\boldsymbol{\ell}} + \boldsymbol{F}_{\boldsymbol{m}_\rho}\right) \cdot \boldsymbol{e}_x \tag{33}$$

$$D_p^d \stackrel{\text{def}}{=} \left(\boldsymbol{F}_{S\mathcal{V}} + \boldsymbol{F}_\mu + \boldsymbol{F}_{\nabla\rho}\right) \cdot \boldsymbol{e}_x \approx \boldsymbol{F}_{S\mathcal{V}} \cdot \boldsymbol{e}_x \tag{34}$$

where, apart from $\boldsymbol{F}_\mu$, the contribution of $\boldsymbol{F}_{\nabla\rho}$ has also been dropped from Eqs. (33) and (34). Indeed an equivalent expression for $\boldsymbol{F}_{\nabla\rho}$ based on volume integrals has also been derived by Minervino and Tognaccini [33]. Nevertheless, as its contribution would approach zero when integrated over the entire flow-field [10], including its local force generation in the visualization of elementary lift-induced and parasite drag can be confusing, therefore only those contributions that are non-globally-zero were included in the contour plots of Figs. 15 and 16. Again a non-zero contribution to lift-induced and parasite drag is only obtained in the rotational flow-field regions.

A similar visualization can also be provided for irreversible drag, as the divergence form of Eq. (6) is suitable at the scope. In Fig. 17, the formulation of Destarac and van der Vooren [7] has been used and, comparing to parasite drag contour plot in Fig. 16, it is clear how the irreversible drag generation is well confined to viscous and shock wave regions, which is less visible in the contour plot of local parasite drag.

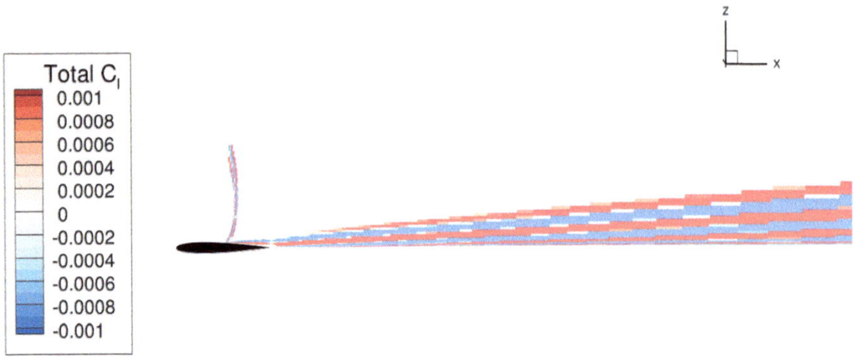

Fig. 13 Volumetric contribution to total lift of computational grid cells, viscous transonic flow-field around NACA0012 airfoil ($M = 0.8$; $\mathrm{Re} = 9 \times 10^6$, $\alpha = 3°$)

Fig. 14 Volumetric contribution to total drag of computational grid cells, viscous transonic flow-field around NACA0012 airfoil ($M = 0.8$; $\mathrm{Re} = 9 \times 10^6$, $\alpha = 3°$)

Fig. 15 Volumetric contribution to lift-induced drag of computational grid cells, viscous transonic flow-field around NACA0012 airfoil ($M = 0.8$; $\mathrm{Re} = 9 \times 10^6$, $\alpha = 3°$), excluding the $\boldsymbol{F}_{\nabla\rho}$ contribution

Fig. 16 Volumetric contribution to parasite drag of computational grid cells, viscous transonic flow-field around NACA0012 airfoil ($M = 0.8$; $\mathrm{Re} = 9 \times 10^6$, $\alpha = 3°$), excluding the $\boldsymbol{F}_{\nabla\rho}$ contribution

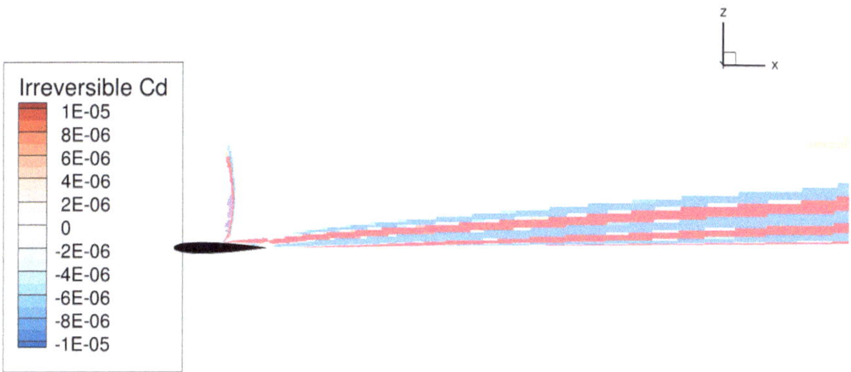

Fig. 17 Volumetric contribution to irreversible drag of computational grid cells, viscous transonic flow-field around NACA0012 airfoil ($M = 0.8$; $\mathrm{Re} = 9 \times 10^6$, $\alpha = 3°$), computed using Destarac and van der Vooren method

5 Conclusions

The employed definitions for the different forms of aerodynamic drag, identified on a physical and phenomenological basis, were first discussed, highlighting their objective or non-objective characteristics in either real or numerically computed flow-fields. The relationship between *lift-induced* and *reversible* drag, as well as between *parasite* and *irreversible* drag, was also theoretically discussed.

Then the *operative* definitions (i.e. the available methods) used to predict the different drag contributions were illustrated, considering classical, thermodynamic-based and Lamb-vector-based approaches. Generalized forms of the Kutta-Joukowski theorem, Maskell and Betz formulas were derived starting from the integral momentum balance, for steady flows, as an alternative to the originally proposed derivation starting from the vortical force formulas. The state-of-the-art of lift-induced and parasite drag definitions in a generally viscous and compressible flow was discussed, highlighting the underlying hypotheses and limitations.

The capability of Lamb-vector-based lift-induced and parasite drag definitions (directly linked to generalized classical formulas) to respectively capture reversible and irreversible drag contributions was theoretically discussed, then numerical support to the theoretical argumentation was provided post-processing three-dimensional and two-dimensional CFD solutions. The latter ones were analyzed in nearly incompressible, subsonic and transonic regime, considering both inviscid and viscous flows (at high or low Reynolds number). In 2D, consistency of lift-induced drag predictions with the expected zero reversible drag (confirmed by thermodynamic methods) was shown to hold on very large control volumes only. A spurious induced drag contribution, related to the axial velocity perturbation, is otherwise computed, resulting in negative induced drag values.

The above argumentation suggests that a drag breakdown in lift-induced and parasite contributions, capturing respectively reversible and irreversible drag components, cannot be obtained with an arbitrary choice of the external integration domain surface (Σ), that should be placed far away from the body surface in order to effectively decompose the total drag in reversible and irreversible parts.

Unfortunately, in numerical applications, the integration domain cannot be extended too far downstream of the body, where a spurious conversion of reversible into irreversible drag also occurs, as discussed in Sect. 2.1 and evidenced in Sect. 4. This limits the accuracy of the drag breakdown achievable through classical or Lamb-vector-based methods, and highlights that the proposed direct definitions of lift-induced drag in viscous and compressible flows still need to be sharpened.

References

1. Betz, A.: A method for the direct determination of wing-section drag. Tech. Rep. NACA TM37, National Advisory Committee for Aeronautics (1925)
2. Bisson, F.: Aerodynamic Design Optimization Using the Drag Decomposition Method. Master's thesis, Department of Mechanical Engineering, McGill University, Montreal, Quebec (2015). https://escholarship.mcgill.ca/concern/theses/wh246v767
3. Bryant, L.W., Williams, D.H.: An investigation of the flow of air an aerofoil of infinite span. Philos. Trans. Roy. Soc. Lond. Ser. A Contain. Pap. Math. Phys. Charact. **225**(626–635), 199–245 (1926). https://doi.org/10.1098/rsta.1926.0005
4. Chan, W.M., Chiu, I.T., Buning, P.G.: User's manual for the HYPGEN hyperbolic grid generator and the HGUI graphical user interface. Technical Memorandum (TM) 19940011311, NASA National Aeronautics and Space Administration, AMES Research Center, Moffett Field,California (1993). https://ntrs.nasa.gov/citations/19940011311
5. Chao, D.D., Van Dam, C.P.: Wing drag prediction and decomposition. J. Aircraft **43**(1), 82–90 (2006). https://doi.org/10.2514/1.12311
6. Cummings, R., Giles, M., Shrinivas, G.: Analysis of the elements of drag in three-dimensional viscous and inviscid flows. In: Proceedings of the 14th Applied Aerodynamics Conference. No. 0 in Fluid Dynamics and Co-located Conferences, American Institute of Aeronautics and Astronautics (1996). https://doi.org/10.2514/6.1996-2482

7. Destarac, D., van der Vooren, J.: Drag/thrust analysis of jet-propelled transonic transport aircraft; definition of physical drag components. Aerosp. Sci. Technol. **8**(6), 545–556 (2004). https://doi.org/10.1016/j.ast.2004.03.004, https://www.sciencedirect.com/science/article/pii/S1270963804000343
8. Drela, M.: Power balance in aerodynamic flows. AIAA J. **47**(7), 1761–1771 (2009). https://doi.org/10.2514/1.42409
9. Fournis, C.: Wave drag prediction using the Lamb vector in two-dimensional steady transonic flows. In: Proceedings of the AIAA AVIATION 2022 Forum. No. 0 in AIAA AVIATION Forum, American Institute of Aeronautics and Astronautics (2022). https://doi.org/10.2514/6.2022-3886
10. Fournis, C., Bailly, D., Tognaccini, R.: Compressibility correction to Kutta-Joukowski and Maskell formulas using vortex-force theory. AIAA J. **59**(2), 773–778 (2021). https://doi.org/10.2514/1.J059710
11. Fournis, C., Bailly, D., Tognaccini, R.: Invariant Vortex-force theory extending classical aerodynamic theories to transonic flows. AIAA J. **60**(9), 5070–5082 (2022). https://doi.org/10.2514/1.J061278
12. Gariepy, M., Trepanier, J.Y., Malouin, B.: Generalization of the far-field drag decomposition method to unsteady flows. AIAA J. **51**(6), 1309–1319 (2013). https://doi.org/10.2514/1.J051609
13. Giles, M.B., Cummings, R.M.: Wake integration for three-dimensional flowfield computations: theoretical development. J. Aircraft **36**(2), 357–365 (1999). https://doi.org/10.2514/2.2465
14. Hunt, D.L., Cummings, R.M., Giles, M.B.: Wake integration for three-dimensional flowfield computations: applications. J. Aircraft **36**(2), 366–373 (1999). https://doi.org/10.2514/2.2466
15. Jameson, A., Schmidt, W., Turkel, E.: Numerical solution of the Euler equations by finite volume methods using Runge Kutta time stepping schemes. In: Proceedings of the 14th Fluid and Plasma Dynamics Conference. No. 0 in Fluid Dynamics and Co-located Conferences, American Institute of Aeronautics and Astronautics (1981). https://doi.org/10.2514/6.1981-1259
16. Joukowski, N.E.: On annexed vortices. Proc. Phys. Sect. Nat. Sci. Soc. **13**(2), 12–25 (1906)
17. Kang, L.L., Russo, L., Tognaccini, R., Wu, J.Z., Su, W.D.: Aerodynamic force breakdown in reversible and irreversible components by vortex force theory. AIAA J. **57**(11), 4623–4638 (2019). https://doi.org/10.2514/1.J058379
18. Kok, J.C.: Resolving the dependence on freestream values for the k-turbulence model. AIAA J. **38**(7), 1292–1295 (2000). https://doi.org/10.2514/2.1101
19. Kusunose, K.: Drag prediction based on a wake-integral method. In: Proceedings of the 16th AIAA Applied Aerodynamics Conference. No. 0 in Fluid Dynamics and Co-located Conferences, American Institute of Aeronautics and Astronautics (1998). https://doi.org/10.2514/6.1998-2723
20. Kutta, W.M.: Auftriebskräfte in Strömenden Flüssigkeiten. In: IllustrierteAeronautische Mitteilungen 133, vol. 6, pp. 133–135 (1902)
21. Liu, L.Q., Wu, J.Z., Shi, Y.P., Zhu, J.Y.: A dynamic counterpart of Lamb vector in viscous compressible aerodynamics. Fluid Dyn. Res. **46**(6), 061417 (2014). https://doi.org/10.1088/0169-5983/46/6/061417
22. Liu, L.Q., Zhu, J.Y., Wu, J.Z.: Lift and drag in two-dimensional steady viscous and compressible flow. J. Fluid Mech. **784**, 304–341 (2015). https://doi.org/10.1017/jfm.2015.584, https://www.cambridge.org/core/article/lift-and-drag-in-twodimensional-steady-viscous-and-compressible-flow/D2D278D4CB0BFA4519BD331B9666540A
23. Liu, L., Shi, Y., Zhu, J., Su, W., Zou, S., Wu, J.: Longitudinal–transverse aerodynamic force in viscous compressible complex flow. J. Fluid Mech. **756**, 226–251 (2014). https://doi.org/10.1017/jfm.2014.403, https://www.cambridge.org/core/journals/journal-of-fluid-mechanics/article/longitudinaltransverse-aerodynamic-force-in-viscous-compressible-complex-flow/599729CF6EA2106557911EB376686188
24. Liu, L., Wu, J.Z., Su, W., Kang, L.: Lift and drag in three-dimensional steady viscous and compressible flow. Phys. Fluids **29** (2017). https://doi.org/10.1063/1.4989747

25. Marongiu, C., Tognaccini, R.: Far-field analysis of the aerodynamic force by Lamb vector integrals. AIAA J. **48**(11), 2543–2555 (2010). https://doi.org/10.2514/1.J050326
26. Marongiu, C., Catalano, P., Amato, M., Iaccarino, G.: U-ZEN: a computational tool solving U-Rans equations for industrial unsteady applications. In: Proceedings of the 34th AIAA Fluid Dynamics Conference and Exhibit. No. 0 in Fluid Dynamics and Co-located Conferences, American Institute of Aeronautics and Astronautics (2004). https://doi.org/10.2514/6.2004-2345
27. Marongiu, C., Tognaccini, R., Ueno, M.: Lift and lift-induced drag computation by Lamb vector integration. AIAA J. **51**(6), 1420–1430 (2013). https://doi.org/10.2514/1.J052104
28. Maskell, E.C.: Progress towards a method for the measurement of the components of the drag of a wing of finite span. Tech. Rep. Technical Report No. 72232, RAE, United Kingdom (1972)
29. Mele, B., Ostieri, M., Tognaccini, R.: Vorticity based breakdown of the aerodynamic force in three-dimensional compressible flows. AIAA J. **54**(4), 1198–1208 (2016). https://doi.org/10.2514/1.J054363
30. Mele, B., Ostieri, M., Tognaccini, R.: Aircraft lift and drag decomposition in transonic flows. J. Aircraft **54**(5), 1933–1944 (2017). https://doi.org/10.2514/1.C034288
31. Mele, B., Tognaccini, R.: Aerodynamic force by Lamb vector integrals in compressible flow. Phys. Fluids **26**(5), 056104 (2014). https://doi.org/10.1063/1.4875015
32. Minervino, M., Tognaccini, R.: A unified thermodynamic/Lamb-vector-based analysis of the aerodynamic force. Phys. Fluids **35**(9), 097135 (2023). https://doi.org/10.1063/5.0164384
33. Minervino, M., Tognaccini, R.: On the spurious effects in Lamb-vector-based force decomposition methods. Aerosp. Sci. Technol. **142**, 108674 (2023). https://doi.org/10.1016/j.ast.2023.108674, https://www.sciencedirect.com/science/article/pii/S1270963823005709
34. Minervino, M., Tognaccini, R.: Wave drag and vorticity in two-dimensional viscous and inviscid flows. In: Proceedings of the AIAA SCITECH 2023 Forum. No. 2023-1032 in AIAA SciTech Forum, American Institute of Aeronautics and Astronautics (2023). https://doi.org/10.2514/6.2023-1032
35. Morkovin, M.V.: Effects of compressibility on turbulent flow. In: The Mechanics of Turbulence, p. 367. Gordon and Breach (1962)
36. Noca, F., Shiels, D., Jeon, D.: A comparison of methods for evaluating time-dependent fluid dynamic forces on bodies, using only velocity fields and their derivatives. J. Fluids Struct. **13**(5), 551–578 (1999). https://doi.org/10.1006/jfls.1999.0219, https://www.sciencedirect.com/science/article/pii/S0889974699902190
37. Ostieri, M., Mele, B., Tognaccini, R.: Linear and nonlinear decomposition of aerodynamic force acting on an oscillating plate. AIAA J. **56**(2), 594–608 (2018). https://doi.org/10.2514/1.J056129
38. Ostieri, M., Tognaccini, R.: On a recently proposed vorticity-based definition of wave drag. J. Aircraft **55**(6), 2521–2523 (2018). https://doi.org/10.2514/1.C034820
39. Oswatitsch, K.: Gas dynamics. Applied Mathematics and Mechanics. Academic (1956). https://books.google.it/books?id=mfZQAAAAMAAJ
40. Paparone, L., Tognaccini, R.: Computational fluid dynamics-based drag prediction and decomposition. AIAA J. **41**(9), 1647–1657 (2003). https://doi.org/10.2514/2.7300
41. Prandtl, L., Tietjens, O.G.: Applied Hydro- and Aeromechanics. Dover Publications (1934)
42. Prandtl, L.: Theory of lifting surfaces, part 1. Technical Report NACA TN-10, National Advisory Committee for Aeronautics (1919)
43. Rumsey, C.: Turbulence Modeling Resource—2D NACA 0012 Airfoil Validation Case. National Aeronautics and Space Administration (2021). https://turbmodels.larc.nasa.gov/naca0012_val.html. Accessed 14 Dec 2023
44. Saffman, P.G.: Vortex Dynamics. Cambridge Monographs on Mechanics, Cambridge University Press (1993). https://doi.org/10.1017/CBO9780511624063, https://www.cambridge.org/core/books/vortex-dynamics/2F3335617172249253B949EA3406F89
45. Schmitz, S.: Drag decomposition using partial-pressure fields in the compressible Navier-Stokes equations. AIAA J. **57**(5), 2030–2038 (Jan2019). https://doi.org/10.2514/1.J057701

46. Schmitz, S., Coder, J.G.: Inviscid circulatory-pressure field derived from the incompressible Navier-Stokes equations. AIAA J. **53**(1), 33–41 (2015). https://doi.org/10.2514/1.J053140
47. Spalart, P.R.: On the far wake and induced drag of aircraft. Journal of Fluid Mechanics **603**, 413–430 (2008). https://doi.org/10.1017/S0022112008001146
48. Tognaccini, R., Saetta, E., Mele, B.: Aerodynamic force breakdown: thermodynamic vs vortex-force methods. In: Proceedings of the AIAA AVIATION 2021 FORUM. No. 0 in AIAA AVIATION Forum, American Institute of Aeronautics and Astronautics (2021). https://doi.org/10.2514/6.2021-2553
49. Toubin, H., Bailly, D.: Development and application of a new unsteady far-field drag decomposition method. AIAA J. **53**(11), 3414–3429 (2015). https://doi.org/10.2514/1.J054002
50. Toubin, H., Bailly, D., Costes, M.: Improved Unsteady Far-Field Drag Breakdown Method and Application to Complex Cases. AIAA Journal **54**(6), 1907–1921 (2016). https://doi.org/10.2514/1.J054756
51. Ueno, M., Yamamoto, K., Tanaka, K., Murayama, M., Tognaccini, R.: Far-field drag analysis of NASA common research model simulation. J. Aircraft **50**(2), 388–397 (2013). https://doi.org/10.2514/1.J05475610.2514/1.C031766
52. van Dam, C.P.: Recent experience with different methods of drag prediction. Prog. Aerosp. Sci. **35**(8), 751–798 (1999). https://doi.org/10.1016/S0376-0421(99)00009-3, https://www.sciencedirect.com/science/article/pii/S0376042199000093
53. van der Vooren, J., Slooff, J.W.: CFD-based drag prediction; state of the art, theory, prospects. Lectures Notes NLR TP 90247, Lectures Notes, AIAA Professional Studies Series, Course on Drag-Prediction and Measurement. Dutch Aerospace Lab. (1990)
54. Wu, J.Z., Lu, X.Y., Zhuang, L.X.: Integral force acting on a body due to local flow structures. J. Fluid Mech. **576**, 265–286 (2007). https://doi.org/10.1017/S0022112006004551, https://www.cambridge.org/core/journals/journal-of-fluid-mechanics/article/integral-force-acting-on-a-body-due-to-local-flow-structures/ECC4554D0C220D877514DBA82B2761AA
55. Wu, J.Z., Ma, H.Y., Zhou, M.D.: Vorticity and Vortex Dynamics. Springer, Berlin, Heidelberg (2006). https://doi.org/10.1007/978-3-540-29028-5
56. Wu, J., Liu, L., Liu, T.: Fundamental theories of aerodynamic force in viscous and compressible complex flows. Prog. Aerosp. Sci. **99**, 27–63 (2018). https://doi.org/10.1016/j.paerosci.2018.04.002, https://www.sciencedirect.com/science/article/pii/S0376042117301355
57. Yamazaki, W., Matsushima, K., Nakahashi, K.: Aerodynamic design optimization using the drag-decomposition method. AIAA J. **46**(5), 1096–1106 (2008). https://doi.org/10.2514/1.30342

Enhancing Aircraft Design and Performance Through High-Fidelity Digital Twins: A CFD-Based Approach

Giovanna Gargiulo, Ubaldo Cella, and Marco E. Biancolini

Abstract The Digital Twin (DT) concept lacks a unique definition, as it varies depending on the application field. A DT embodies the virtualization of a physical asset and plays a vital role in real-time prediction, optimization, monitoring, control, and informed decision-making. It enables reactive, preventive, predictive, and prescriptive maintenance while facilitating efficient data processing and rapid simulations. In the aerospace industry, creating high-fidelity models is crucial for enhancing new aircraft's preliminary design, performance, and safety. A digital twin can support shifting from analytical methods to estimate stability and element dimensions, addressing challenges like digital transformation, high production operation costs, low maintenance efficiency, and intensive technology (Xiong and Wang in Int J Adv Manuf Technol 121:5677–5692, 2022 [1]). The Next Generation Air Dominance encompasses high-fidelity physical models and avionics software, requiring a digital representation. West et al. (Procedia Comput Sci 114:47–56, 2017 [2]) provides insights into establishing a platform for digital twinning in the aerospace industry. Improving models to construct high-fidelity digital twins is imperative. Despite utilizing significant computational resources, such as Computer Fluid Dynamics (CFD), the Reduced Order Model (ROM) enables accurate and swift simulations of factors like aerodynamic loads. (CFD based ROMs for aeronautical applications, 2016 [3]). Essentially, the digital twin can be employed to validate the data required to estimate the dynamic performance of aircraft. This study is interested in a glider scenario resembling a game simulation. By conducting CFD simulations, enough variables of interest are obtained, enabling the creation of a ROM and, subsequently, the digital twin. This approach showcases the advantages of employing the ROM and digital twin for evaluating aircraft performance and aerodynamic loads.

Keywords Glider · CFD · ROM · Digital twin

G. Gargiulo · U. Cella · M. E. Biancolini (✉)
Department of Enterprise Engineering, Faculty of Engineering, University of Rome Tor Vergata, Rome, Italy
e-mail: biancolini@ing.uniroma2.it

M. E. Biancolini et al. (eds.), *Digital Twins for Aerospace*, Springer Aerospace Technology, https://doi.org/10.1007/978-3-032-05022-9_5

1 Introduction

Using a DT and virtualizing physical assets facilitates real-time prediction, optimization, monitoring, control, and improved decision-making through the rapid processing of data and simulation execution. Rasheed et al. [4] describe the DT as a computational metamodel, device shadow, mirrored system, avatar, or synchronized virtual prototype. The authors also introduce the concept of digital siblings, copies of physical assets, which are useful for testing hypothetical scenarios. Furthermore, they outline eight properties of a digital twin and provide examples of application areas:

1. Real-time remote monitoring and control
2. Greater efficiency and safety
3. Predictive maintenance and scheduling
4. Scenario and risk assessment
5. Better intra- and inter-team synergy and collaborations
6. More efficient and informed decision support system
7. Personalization of products and services
8. Better documentation and communication.

The main areas of interest encompass Meteorology, Manufacturing and process technology, and Education.

The "manufacturing and process technology" sector is particularly relevant for employing DTs, with studies by VanDerHorn and Mahadevan [5] and Tao et al. [6] providing valuable insights; further, Psarommatis and May [7] conducted a comprehensive review of this sector.

Liu et al. [8] define a DT as an evolution of simple Computer-Aided Design (CAD) and sensor-based Internet of Things (IoT). This evolution implies enhanced interactions between different components and life cycle processes, increasing interaction capabilities between the digital and physical worlds, thereby improving artificial intelligence (AI) models and smart manufacturing profiles, Dreyfus et al. [9].

Bachelor et al. [10] about the research project CRYSTAL, explore lifecycle and non-life cycle-related interoperability aspects, focusing on standards connection and heterogeneous simulation. They illustrate their findings by designing an ice protection system for a regional aircraft. In particular, the aviation sector faces challenges, including digital transformation, high production operation costs, low maintenance efficiency, and intensive technology, Xiong and Wang [1]. In response, a methodology called 'Certification-by-Simulation' is proposed, emphasizing the adoption of simulation over real campaign tests where possible. The NASA Vision 2030 aims to demonstrate compliance with airworthiness requirements.

The DT introduces a groundbreaking approach by constructing high-fidelity and Reduced ROM models. The build of the ROM, in the initial step, leverages massive CFD computational resources and optimizes computational costs in subsequent steps. At the end of the process, we obtain our interest results within minutes. [3], Fig. 1a. We can apply this technique and obtain the data essential for estimating the dynamic performances of aircraft.

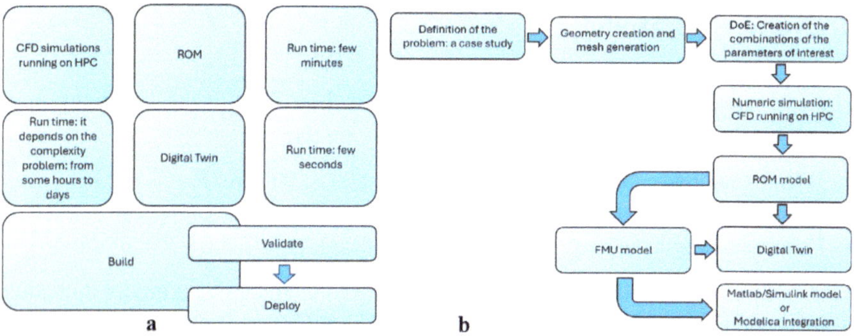

Fig. 1 **a** Time scheme of different computational elements used to build the DT; **b** Workflow: from the case study to FMU/TB

In previous aeronautical studies, Tuegel et al. [11] presented a DT predicting the life of aircraft structures to ensure structural integrity, integrating local damages, and considering structural deflections and temperature in flight conditions. Laukotka and Krause [12] focused on implementing DT for handling available data, planning and executing design, and adapting new cabins.

Building on the work of Biancolini et al. [13], where a glider in the Red Bull competition was considered, the current study explores a glider scenario that could be similar to a game simulation. The idea is to generate a flight dynamics model to be linked to a flight simulation architecture. Several tools and strategies are possible to this aim. In Berndt and De Marco [14], the description of the open source JSBSim flight dynamics model, suitable to be fed by a custom aircraft models, is reported. The libraries can be linked to architectures such as FlightGear or OpenEaagles. The workflow here presented is based on the generation of a database of high fidelity CFD solutions and the generation of a ROM for a Functional Mock-up Unit (FMU) model which will be imported into Simulink (see Fig. 1b). Different solutions with the Modelica language are currently being studied. By providing the appropriate input variables and modifying the number of output variables in the FMU model, it could be possible to connect it with the FlightGear software, partially following the approach suggested by Tamayo et al. [15], Cannon et al. [16] and Turvesky et al. [17]. Portapas et al. [18], realized an integration between MATLAB/Simulink environment for flight dynamics modelling with the AGI STK Aviator mission simulator. Generally employing Matlab to create a management system for the aircraft and connecting the joystick Lum in [19], as demonstrated in different works, to lead the glider is viable. ROM models are exported first to Ansys Twin Builder and, subsequently, for the appropriate input variables, to Matlab/Simulink. To achieve this, we start with CFD simulations, from which various fields of interest variables are obtained to build the ROM and, subsequently, the DT.

2 Numerical Model

The case study considers an hypothetical glider, Fig. 2, with the following main dimensions:

- **Wing**: span bw = 8.34 m, area Aw = 8.72 m^2, taper ratio TRw = 0.13
- **Horizontal tail**: span bh = 2.5 m, area Ah = 2.27 m^2
- **Fuselage**: length fuselage Lf = 9.54 m.

This study is interested in an initial investigation to characterize a model that could be integrated into a flight simulator such as the FlightGear software. The operational conditions consider a starting altitude of 500 m above sea level and a velocity of around 100 m/s (a velocity higher than the usual one used to enlarge the training range for the successive ROM), which implied an $M < 0.3$. Equations 2.1 and 2.2 provide the definition for the lift and drag coefficients. Equation 2.3 is the expression of the pitching coefficients.

$$C_L = \frac{2L}{\rho U^2 S} \tag{2.1}$$

$$C_D = \frac{2D}{\rho U^2 S} \tag{2.2}$$

$$C_M = \frac{2M}{\rho U^2 S \, MAC} \tag{2.3}$$

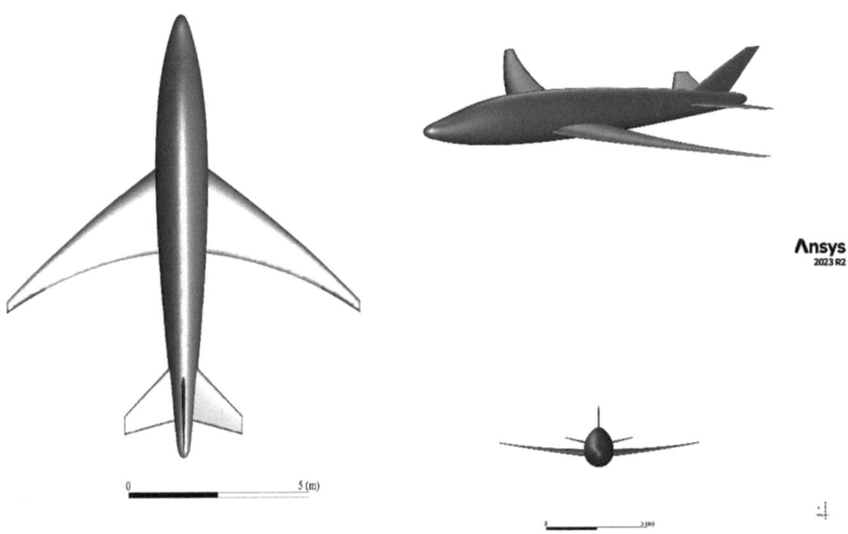

Fig. 2 View of the adopted aircraft model

The aerodynamic efficiency is defined by Eq. 2.4, and determine the glide ratio. The flight range is expressed in Eq. 2.5. These calculations offer crucial metrics for evaluating the aerodynamic performance and capabilities of the glider in terms of height and range during flight.

$$\mathrm{E} = \frac{\mathrm{L}}{\mathrm{D}} \tag{2.4}$$

$$E = \frac{\mathrm{h}}{\mathrm{range}} \tag{2.5}$$

2.1 Computational Model and Mesh

The simulations were conducted on a half model, assuming the symmetry of the flow field due to the absence of a sideslip. The numerical domain consisted in an unstructured tetrahedral mesh with about half million of cells. The analyses were executed using Ansys Fluent V23r2, integrating parametric analysis with RBF Morph Fluids and Twin Builder (TB). Steady incompressible RANS (Reynolds Average Navier Stokes) analyses using a Pressure-Based Solver were run. Second-order spatial discretization was adopted for continuity, and momentum equations. Turbulence was modeled using a k-ε model with a standard wall function.

The ROM was built from CFD simulations. We selected four input parameters, one physical parameter (velocity in the far field), and three geometrical parameters (the angle of the elevator, flaps, and trims) for the simulations. These parameters of interest (Table 1) are spanned the range between the minimum and maximum possible values. For a correct number of combinations to construct the ROM, we need a minimum of 2^3 that will be multiplied by the four parameters used in the simulations. We defined a DoE (Design of Experiments) table with 256 combinations of parameters selected using an "Optimal Space-Filling Design" methodology and incorporating both the maximum and minimum parameter values. This approach ensures a well-distributed coverage in the parameter space, facilitating a comprehensive field exploration. The ROM was subsequently exported as FMU and imported into TB or directly imported into TB.

Table 1 The range of variation of parameters

Parameters	Min values	Max values
Far field [m/s]	10	30
Elevator [α]	– 15.0	15.0
Flaps [α]	0.0	30.0
Trim [α]	– 5.0	5.0

Morphing Concept

The morphing of the geometry through Radial Basis Function (RBF), as described by Biancolini [20], involves applying a displacement to a set of points, specifically the nodes (target points) of the computational model. This displacement is then interpolated. The choice of radial function and its support influences the interpolation outcome.

RBF computes unknown weight functions and polynomial terms based on known data. Equation 2.6 [20] outlines the general form of the interpolation function and the polynomial:

$$s(\boldsymbol{x}) = \sum_{i=1}^{N} \gamma_i \varphi\left(\|\boldsymbol{x} - \boldsymbol{x}_{ki}\|\right) + h(\boldsymbol{x}) \tag{2.6}$$

The scalar function s(x) is defined for each arbitrary point in space x and represents a transformation $\mathbb{R}^n \rightarrow \mathbb{R}$. The displacement of a point can be conceptualized as the Euclidean distance between the source $\boldsymbol{x}_{ki}$ and target points x multiplied by the radial function φ and the weight γ_i. The polynomial h(**x**) is then introduced, and its minimum degree depends on the choice of the basis function, as illustrated in Eq. 2.7.

$$h(\boldsymbol{x}) = \beta_1 + \beta_2 x + \beta_3 y + \beta_4 z \tag{2.7}$$

$$s(\boldsymbol{x}_{ki}) = g_i, \quad 1 \leq i \ \leq N \tag{2.8}$$

The desired function values g_s are obtained at source points, as expressed in Eq. 2.8.

The system is considered complete when orthogonality conditions are satisfied for all polynomials pp with a degree less than or equal to that of the polynomial h, as outlined in Eq. 2.9 [20].

$$\sum_{i=1}^{N} \gamma_i h(\boldsymbol{x}_{ki}) = 0 \tag{2.9}$$

In Eqs. 2.10 to 2.12, the interpolation is illustrated for a 3D set of displacements at source points.

$$s_x(\boldsymbol{x}) = \sum_{i=1}^{N} \gamma_i^x \varphi\left(\|\boldsymbol{x} - \boldsymbol{x}_{ki}\| \right) + \beta_1^x + \beta_2^x x + \beta_3^x y + \beta_4^x z. \tag{2.10}$$

$$s_y(\boldsymbol{x}) = \sum_{i=1}^{N} \gamma_i^y \varphi\left(\|\boldsymbol{x} - \boldsymbol{x}_{ki}\| \right) + \beta_1^y + \beta_2^y x + \beta_3^y y + \beta_4^y z. \tag{2.11}$$

$$s_z(\boldsymbol{x}) = \sum_{i=1}^{N} \gamma_i^z \varphi\left(\|\boldsymbol{x} - \boldsymbol{x}_{ki}\| \right) + \beta_1^z + \beta_2^z x + \beta_3^z y + \beta_4^z z. \tag{2.12}$$

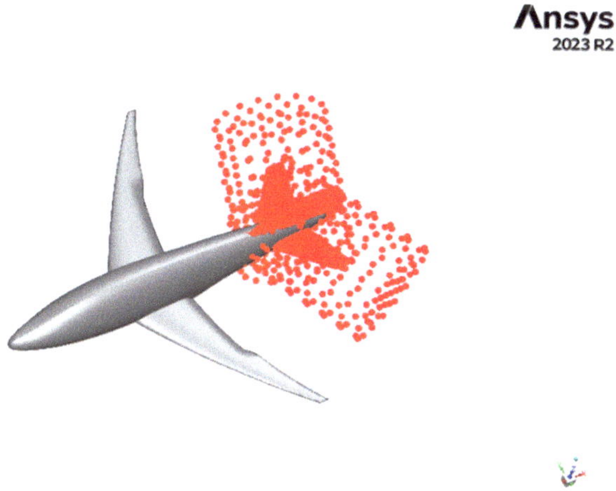

Fig. 3 A view of the glider that exhibits the source points preview delimiting the domain of morphing action to the elevator zone and shows the displacements of flaps

Figure 3 displays the RBF setup adopted for the morphing of the flap. The morphing action is limited to cylindrical regions enveloping the volume around the flap.

Parametric Analyses
A parametric study was conducted considering one physical parameter (far field velocity) and three geometric parameters (elevator, flaps, trims). Table 1 illustrates the range of variations for each parameter. The variables investigated included velocity, pressure, lift, drag, and pitching momentum, which are parameters that could provide valuable insights into the behaviour and performance of the glider under different conditions.

2.2 ROM Construction

After obtaining the snapshots of the field through the parameter combinations defined by the Design of Experiments (Figs. 4 and 5 report the reference solutions), the TB software from Ansys v23r2 was employed to construct the ROM. The process involves compressing the solutions using a Singular Value Decomposition (SVD) algorithm. The superposition of the modes obtained by SVD is then utilized to interpolate the snapshots, enabling the extraction of real-time values for variables of interest at each point within the chosen parametric interval. This approach streamlines the representation of complex system behaviours and allows for efficient interpolation across the parameter space.

The SVD theory considers a rectangular matrix (**M**) that can be decomposed into three matrices, Weiss [21]: two orthogonal matrices (**P**, **Q**) and one diagonal matrix ($\boldsymbol{\Sigma}$), as represented in Eq. 2.13.

$$\mathbf{M} = \mathbf{P}\Sigma Q^T \tag{2.13}$$

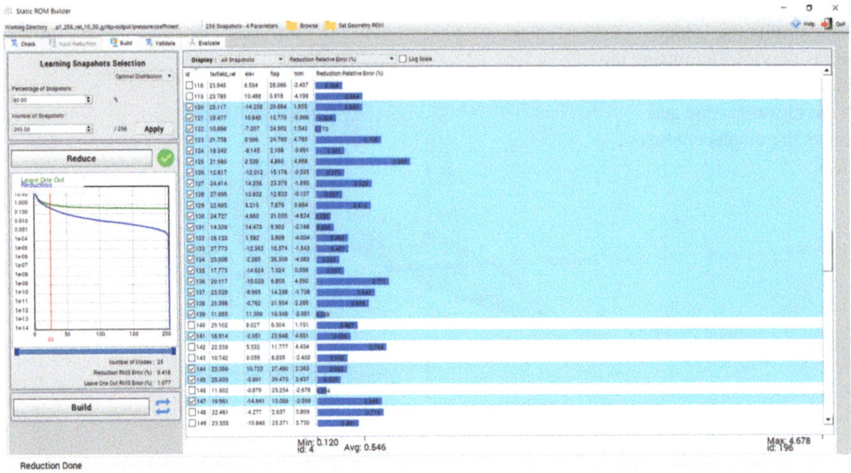

Fig. 4 Pressure coefficient reduction relative error considering the 85% of snapshots

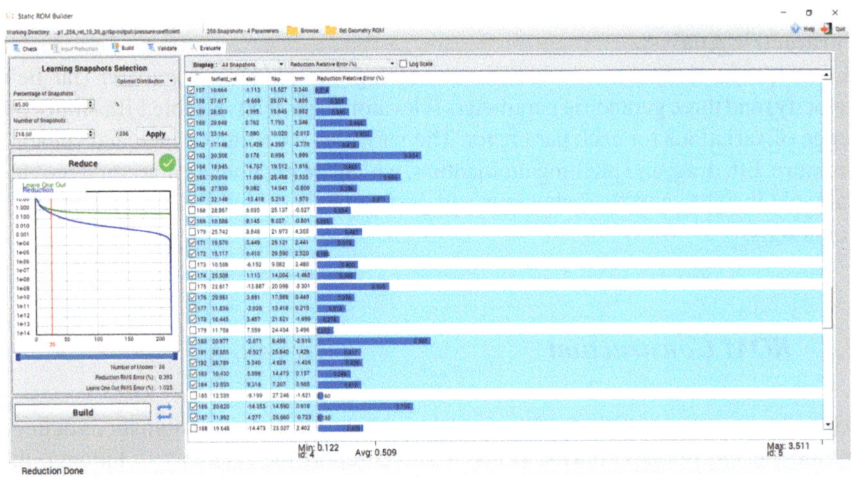

Fig. 5 Pressure coefficient reduction relative error considering the 80% of snapshots

The Singular Value Decomposition (SVD) properties dictate that the matrix M can be approximated as a linear combination of r modes, specifically the first r left singular vectors, with a specific accuracy. The solution field **X** in the design space is expressed by Eq. 2.14:

$$\mathbf{X} = \sum_{i=1}^{r} \alpha_i P_i \tag{2.14}$$

The selection of the number of modes (r) is crucial in constructing the ROM. This number should strike a balance between a larger value for an accurate representation of the data and a smaller value to compress the data while excluding noise. The projection error quickly drops to negligible levels already at small r values.

A response surface of mode coefficients (α_i), defined by an interpolation obtained using the learning snapshots physical and geometrical parameters, is essential for constructing the ROM. It allows to have a direct inference from the design space (input parameters) and the reduced spaces (mode coefficients α_i).

Each new unseen dataset can be expressed through a linear combination of the retained modes weighted by a set of scalar coefficients inferred from the new input parameters using the genetic aggregation response surface (GARS) technique, Ben Salem [22]. The variable v, as defined in Eq. 2.15, is considered as the aggregation of the product of the response surface factor and the mode data for all the modes r:

$$v(\mathbf{x}) = \sum_{i=1}^{r} \text{GARS}_i(\mathbf{x})\ \text{mode}_i(\mathbf{x}) \tag{2.15}$$

To construct the ROM, we partition the snapshots, the instantaneous CFD field for a given set of parameters, into three sets: a learning set (used to build a basis of modes for expressing each solution as a linear combination of these modes), an instantaneous set, and a validation set. Different learning sets can be employed, obtained by selecting the same number of snapshots but considering previously excluded snapshots or by varying the number in the learning set. These variations lead to different errors in terms of relative or absolute errors.

The key errors considered in ROM construction include the relative error of reduction, relative ROM error, maximum absolute error, maximum absolute ROM error, and interpolation error.

Specifically:

- **Interpolation error**: This error is contingent on the interpolation of the mode coefficients.
- **Relative error of reduction**: This error is influenced by the number of modes. It quantifies the difference between the reference solution of each field snapshot and the projected one. In other words, it represents the projection of the reference solution onto the reduced base defined by the number of modes.
- **Relative ROM error**: This error encompasses both the reduction and interpolation errors.

After evaluating the desired error and constructing the ROM, it is possible to export it in TB and assess the evolution of parameters of interest (velocity, pressure, lift, or drag) in the quasi-steady state system. Figures 4 and 5, present the pressure coefficient results regarding relative error of reduction. In Fig. 4, 85% of snapshots are considered, and in Fig. 5, 80% are chosen.

Figures 6 and 7 exhibit the ROM relative error for the pressure coefficient results, they include both reduction and both interpolation errors. Figure 6, 85% of snapshots considers, and Fig. 7, chose the 80% of snapshots. Also, in this case, we observe that the error increases when we adopt fewer learning snapshots.

Adopting a congruent number of snapshots improves accuracy, resulting in more precise predictions, and when we double the number of snapshots, the error reduces.

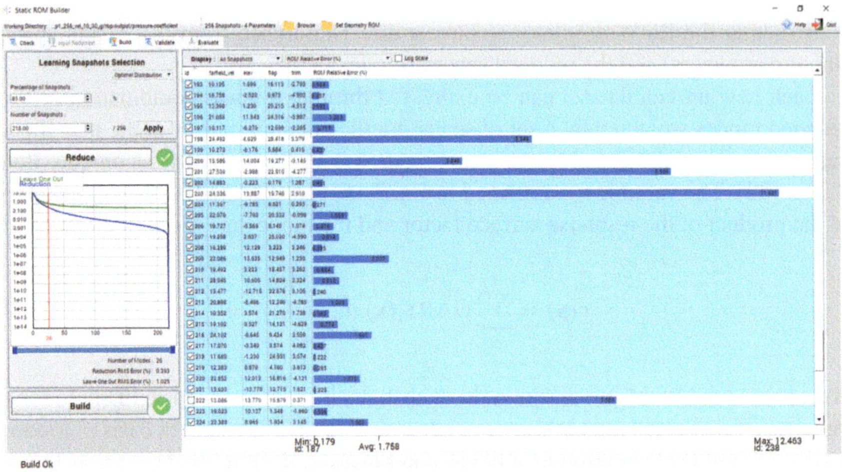

Fig. 6 Pressure coefficient ROM error considering the 85% of snapshots

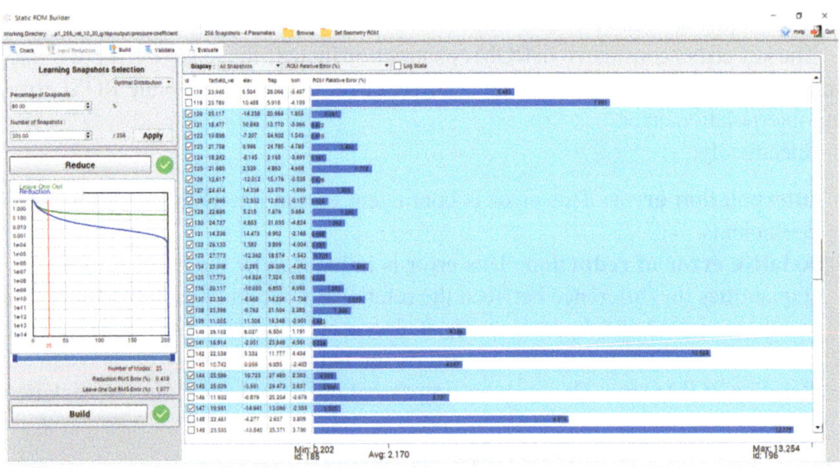

Fig. 7 Pressure coefficient ROM error considering the 80% of snapshots

3 Results and Discussion

In Fig. 8a, a comparison of the second snapshot of two different pressure coefficient ROMs is presented. They are obtained using 80% and 85% of snapshots. By using wall functions in the CFD simulations, it is possible to illustrate in Fig. 8b a comparison between the first snapshots of the turbulent kinetic energy ROM. In both cases, the differences are exhibited in the images on the right side of the figures. It is also possible to evaluate the impact of the modes on the snapshots and subsequent ROM results.

The impact of the combination of ROM and geometric morphing on the flaps (ranging from 0.9 to 24.4) is showed in Figs. 9 and 10 concerning the turbulent kinetic energy field.

The CFD results are compared with the ROM results in Fig. 11. The left figure (Fig. 11a) reports the turbulent kinetic energy field computed by CFD while the right one (Fig. 11b) displays the turbulent kinetic energy estimated by the ROM. The maximum turbulent kinetic energy in CFD is 1.87 m^2/s^2, and in the ROM the maximum value is 2.09 m^2/s^2. The percentage error is around the 12%.

Figures 12 and 13 present the ROM results exported to TB regarding the pressure coefficient and turbulent kinetic energy fields, respectively. The two figures show the field average for output parameters like turbulent kinetic energy and pressure coefficient in two cases: in one case, we consider all components (wing, fuselage, horizontal and vertical plane), and in another, only for one component, the wings. Figure 12 shows the average field of pressure coefficient of all components with a green line while a red line represents the one for the wings component; Fig. 13 shows the average field of turbulent kinetic energy field of all components with a red line while a green line represents the one for the wings component. These visualizations provide insights into the behaviour and characteristics of the glider under different conditions as simulated by the ROM.

This comprehensive analysis provides insights into the overall efficiency and performance of the glider in different configurations and conditions. Additionally, in preparation for future and more comprehensive simulations, the moment for the three axes is obtained and later integrated into the DT simulation software. By changing the parameters listed in Table 1, one can observe the field's evolution. After being imported into Twin Builder, these results can be exported as an FMU and imported into Matlab/Simulink [23] for additional analysis and simulation purposes.

Figure 14a illustrates the *.bat file utilized to execute FlightGear from Simulink. In Figure. 14b, the top image showcases the imported FMU in Simulink for velocity. The bottom image displays the Simulink 3Degree of Freedom (3DoF) block, utilizing three input parameters and accommodating up to six output parameters. By making adjustments to the FMU model, it becomes feasible to include the necessary parameters. Since the FMU model supports only one variable in output, various solutions can be explored, such as modifying the FMU output parameters, employing multiple FMUs, or integrating external parameters.

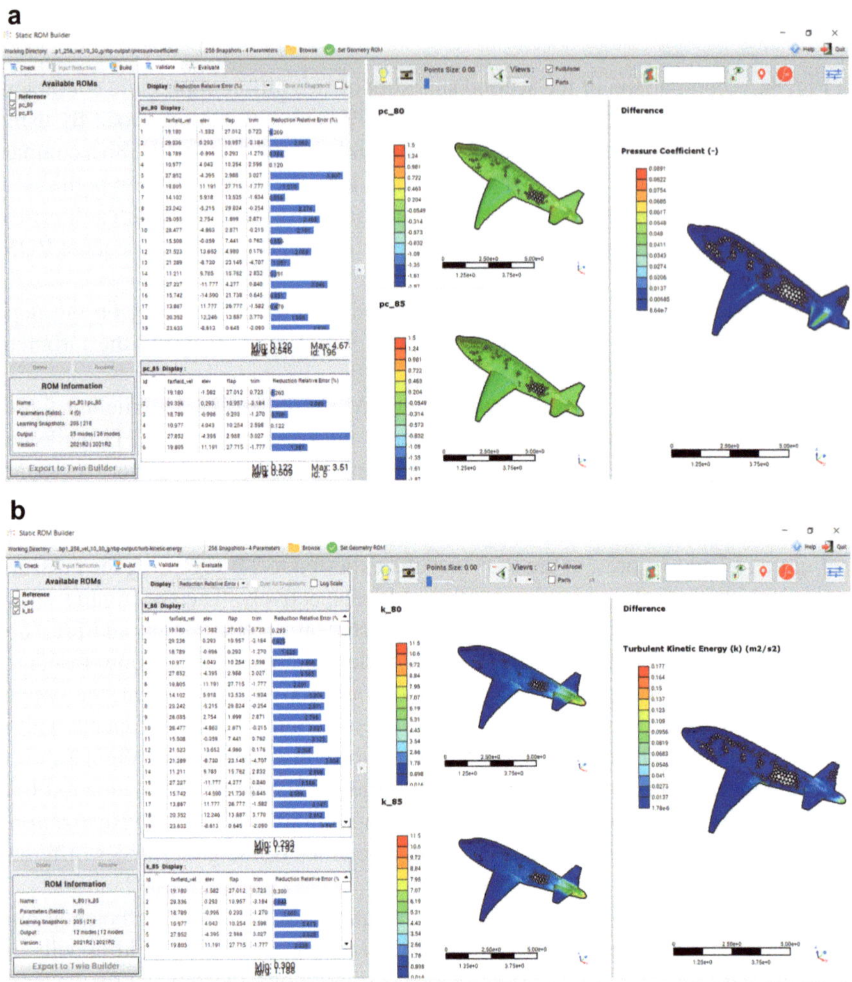

Fig. 8 **a** Pressure coefficient comparison between two different relative reduction errors on the same snapshot but for two different ROMs trained with 80% and 85% snapshots. **b** Turbulent kinetic energy comparison between two different relative reduction errors on the same snapshot but for two different ROMs trained with 80% and 85% snapshots

4 Conclusion

In conclusion, the work has considered generating a database of high-fidelity CFD solutions to generate an ROM and, consequently, to build a DT and an FMU model. The FMU has been imported into Simulink by providing the appropriate input variables and modifying the number of output variables in the FMU model; this model could be connected with the FlightGear software.

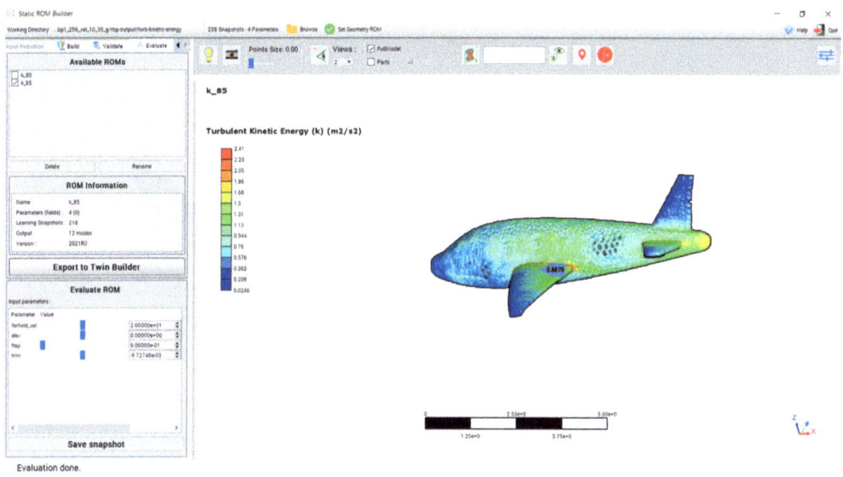

Fig. 9 The influence of the flap angle variation to 0.9

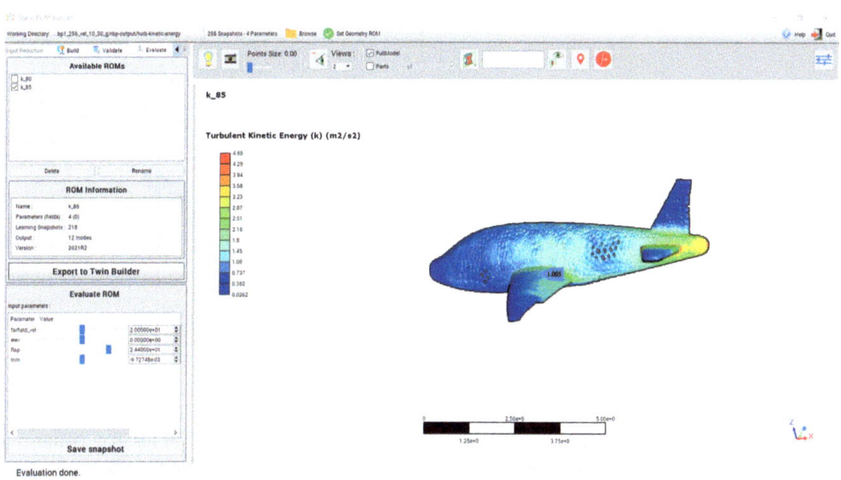

Fig. 10 The influence of the flap angle variation to 24.4

The integration of results into a flight simulation platform similar to FlightGear could enable a dynamic exploration of the glider's behaviour under varying conditions within a simulation environment. This inclusion will allow for a comprehensive and interactive analysis of the glider's performance, enhancing understanding of its dynamics and responsiveness across different scenarios. By exploiting the interaction between Simulink and FlightGear, similar to models realized by Cannon et al. [15] and Turvesky et al. [17], as well as the model for the HL-20 personnel launching system [15] and referencing [24], it will be possible to integrate the glider into a flight

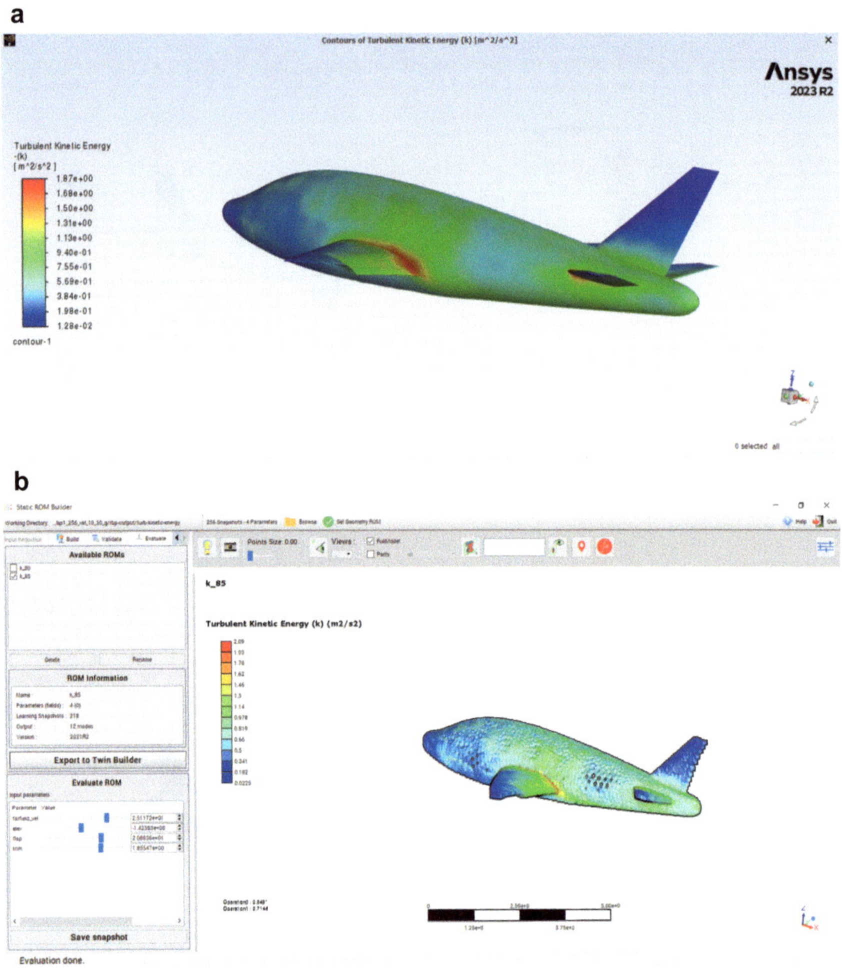

Fig. 11 **a** The turbulent kinetic energy field was obtained with the CFD for the parameter combination obtained from of the snapshot n°120. **b** ROM of the turbulent kinetic energy field

simulator and connect it to an external command [19], such as a joystick, for flight control.

Fig. 15a, b, illustrates respectively a scenario in FlightGear and a scenario that consider the HL-20 example [24]; we show a first implementation of the glider to integrate in the simulator.

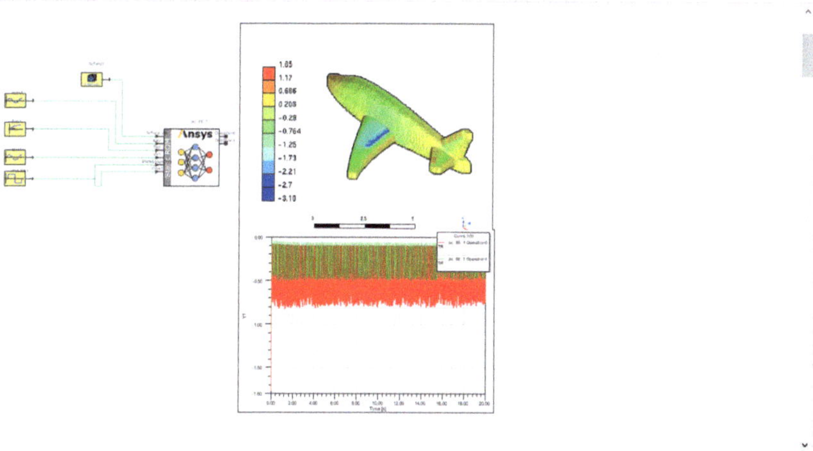

Fig. 12 The DT interface: the left part shows the ROM exported built with the four inputs and one output (the average field), and the right part shows the average field of the pressure coefficient for all components (green line) and for only the wings (red line)

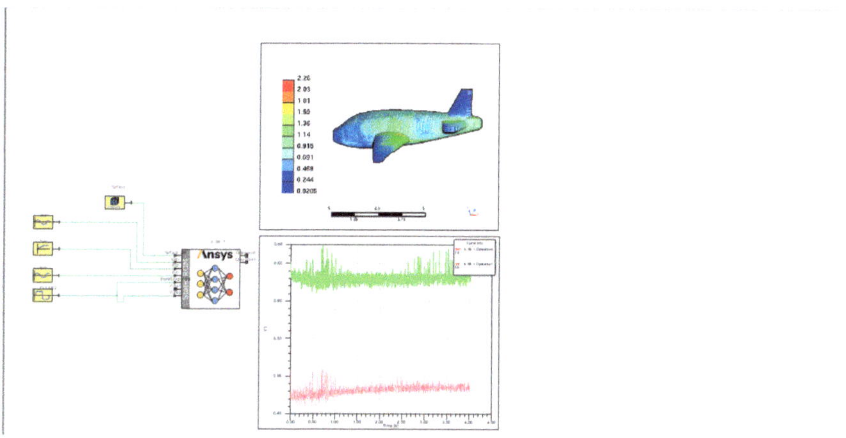

Fig. 13 The DT interface: the left part shows the ROM exported built with the four inputs and one output (the average field), and the right part shows the average field of the turbulent kinetic energy for all components (green line) and for only the wings (red line)

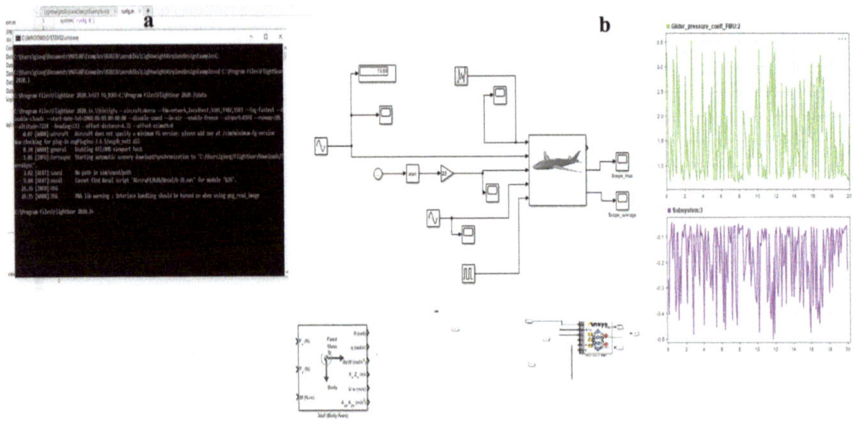

Fig. 14 **a** The *.bat file to integrate Matlab/Simulink with FlightGear. **b** An example of integration of FMU in Simulink

Fig. 15 **a** The figure has been reprinted from FlightGear. **b** The scenario has been reprinted from FlightGear HL-20 simulation

References

1. Xiong, M., Wang, H.: Digital twin applications in aviation industry: a review. Int. J. Adv. Manuf. Technol. **121**(9–10), 5677–5692 (2022)
2. West, T.D., Blackburn, M.: Is digital thread/digital twin affordable? A systemic assessment of the cost of DoD's latest Manhattan project. Procedia Comput. Sci. **114**, 47–56 (2017)
3. Ripepi, M., Karcher, N., Franz, T., Abu-Zurayk, M., Görtz, S.: CFD Based ROMs for Aeronautical Applications (2016)
4. Rasheed, A., San, O., Kvamsdal, T.: Digital twin: values, challenges and enablers from a modeling perspective. IEEE Access **8**, 21980–22012 (2020)

5. VanDerHorn, E., Mahadevan, S.: Digital twin: generalization, characterization and implementation. Decis. Support. Syst. **145**, 113524 (2021)
6. Tao, F., Xiao, B., Qi, Q., Cheng, J., Ji, P.: Digital twin modeling. J. Manuf. Syst. **64**, 372–389 (2022)
7. Psarommatis, F., May, G.: A literature review and design methodology for digital twins in the era of zero-defect manufacturing. Int. J. Prod. Res. **61**(16), 5723–5743 (2023)
8. Liu, J., Zhao, P., Jing, X., Cao, X., Sheng, S., Zhou, H., et al.: Dynamic design method of digital twin process model driven by knowledge-evolution machining features. Int. J. Prod. Res. **60**(7), 2312–2330 (2022)
9. Dreyfus, P.A., Psarommatis, F., May, G., Kiritsis, D.: Virtual metrology as an approach for product quality estimation in industry 4.0: a systematic review and integrative conceptual framework. Int. J. Prod. Res. **60**(2), 742–765 (2022)
10. Bachelor, G., Brusa, E., Ferretto, D., Mitschke, A.: Model-based design of complex aeronautical systems through digital twin and thread concepts. IEEE Syst. J. **14**(2), 1568–1579 (2019)
11. Tuegel, E.J., Ingraffea, A.R., Eason, T.G., Spottswood, S.M.: Reengineering aircraft structural life prediction using a digital twin. Int. J. Aerosp. Eng. **2011** (2011)
12. Laukotka, F.N., Krause, D.: Supporting digital twins for the retrofit in aviation by a model-driven data handling. Systems **11**(3), 142 (2023)
13. Biancolini, M.E., Brutti, C., Porziani, S.: Redbull flugtag: una applicazione "volante" per il cartone ondulato (2019)
14. Berndt, J.S., De Marco, A.: Progress on and usage of the open source flight dynamics model software library JSBSim. In: AIAA Modeling and Simulation Technologies Conference, Chicago, Illinois (2009)
15. Tamayo, S., Gage, S., Walker, G.: Integrated project management tool for modeling and simulation of complex systems. In: AIAA Modeling and Simulation Technologies Conference, p. 4714 (2012)
16. Cannon, M., Gabbard, M., Meyer, T., Morrison, S., Skocik, M., Woods, D.: Swineworks d-200 sky hogg design proposal. AIAA/General Dynamics Corporation Team Aircraft Design Competition (1992)
17. Turevskiy, A., Gage, S., Buhr, C.: Model-based design of a new light-weight aircraft. In: AIAA Modeling and Simulation Technologies Conference and Exhibit, p. 6371 (2007)
18. Portapas, V., Zaidi, Y.: Framework for Developing Digital Twin Prototypes, NATO (2022)
19. Lum, C.: https://www.youtube.com/watch?app=desktop&v=f8tdTiuj5lo, last accessed 03/01/2024 (2020)
20. Biancolini, M.E.: Fast Radial Basis Functions for Engineering Applications. Springer International Publishing, Cham (2017)
21. Weiss, J.: A tutorial on the proper orthogonal decomposition. In: AIAA Aviation 2019 Forum, p. 3333 (2019)
22. Ben Salem, M., Roustant, O., Gamboa, F., Tomaso, L.: Universal prediction distribution for surrogate models. SIAM/ASA J. Uncertain. Quantif. **5**(1), 1086–1109 (2017)
23. Matlab Help 2023
24. https://it.mathworks.com/help/aeroblks/working-with-the-flight-simulator-interface.html#f3-19773, last accessed 03/01/2024

An Interactive Design Tool for NACA Intakes Based on High Fidelity CFD Simulations and Reduced Order Models

Andrea Lopez, Viola Rossano, Ubaldo Cella, Corrado Groth, and Marco E. Biancolini

Abstract The increasing demand for engineers' design capabilities motivates the enhancement of numerical tools to leverage the advancements in computing capabilities. In this context, advanced tools capable of interactively evaluating designs by adjusting component shapes can make a significant difference. The growing adoption of physics-informed artificial intelligence (AI) appears to be a promising avenue. In this paper, we present an interactive dashboard for designing a NACA (National Advisory Committee for Aeronautics) intake, applied to an aeronautical device. The objective was to identify the geometry that maximizes outlet pressure while optimizing pressure recovery. The workflow began with defining 8 design parameters and 1 physical parameter (inlet velocity), applied directly to a high-fidelity CFD model. Shape parameterization was achieved using Radial Basis Functions (RBF) mesh morphing. Subsequently, 80 snapshots were computed using high-performance computing (HPC) and then compressed using Proper Orthogonal Decomposition (POD). Machine learning algorithms facilitated linking the design parameters to mode coefficients. The resulting Reduced Order Model (ROM) enabled real-time prediction of intake performance and visualization of pressure distribution and flow field. An advanced design dashboard was generated, allowing for the selection of a geometry capable of increasing outlet pressure by 160% compared to a baseline intake designed using preliminary design tools.

1 Introduction

The design and optimization of air intakes are crucial aspects of aircraft design, particularly with the increasing electrification of propulsion and the need to dissipate high heat loads. In the aeronautical field, various solutions are available, with two

A. Lopez (✉) · U. Cella · C. Groth · M. E. Biancolini
Enterprise Engineering Department, University of Rome "Tor Vergata", Rome, Italy
e-mail: Lopez@ing.uniroma2.it

V. Rossano
Leonardo Labs, Milan, Italy

M. E. Biancolini et al. (eds.), *Digital Twins for Aerospace*, Springer Aerospace Technology, https://doi.org/10.1007/978-3-032-05022-9_6

commonly employed designs being the NACA air intake and the scoop air intake. The NACA air intake, characterized by its low-profile structure integrated into the aircraft's surface, aims to minimize aerodynamic drag and ensure smooth airflow, thus enhancing overall aerodynamic efficiency. Conversely, the scoop air intake features a more prominent, protruding structure typically positioned externally on the vehicle or aircraft. Its primary function is to capture and direct a larger volume of air into the engine or cooling system, thereby improving engine performance and cooling capabilities. However, the protruding nature of the scoop intake may introduce some degree of drag and turbulence, potentially compromising aerodynamic efficiency.

For electric aircraft, autonomy and low drag are critical factors. Therefore, the focus has been on optimizing NACA air intakes to meet required flow rates and minimize pressure drop (Figs. 1 and 2).

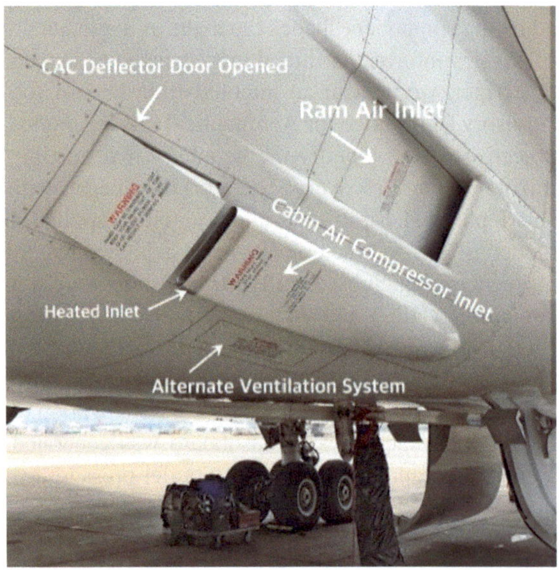

Fig. 1 Two types of air intakes installed on the 787

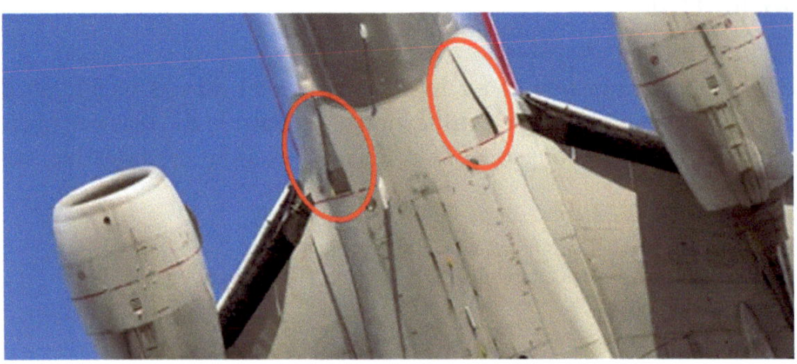

Fig. 2 NACA air intakes

The conventional method for NACA inlet design typically relies on the ESDU standard [1, 2]. This standard, grounded in semi-empirical laws, facilitates the evaluation of optimal configurations based on operating conditions. However, its limitation lies in the narrow range of cases considered in formulating the laws, rendering the assessment of the optimum somewhat unreliable as it pertains solely to the tested scenarios and parameters. In contrast, employing CFD coupled with re-meshing [3] allows for exploration across a broader parameter space, comprehensive case history analysis, and consequently, the identification of superior designs. In this study, an RBF-based morphing technique was adopted to define shape parameters and update the mesh. This approach eliminates meshing noise, reduces computation time, and is well-suited for optimization algorithms [4–8]. Moreover, it maintains mesh topology integrity, facilitating the extraction of Design of Experiment (DOE) data for ROM construction. Leveraging the ROM, a digital twin can be developed for real-time interaction [9–11], a methodology successfully applied across various domains, including biomedical [12], aerospace [13], and multiphysics analysis [14]. In contrast to previous cases, this paper proposes utilizing the ROM to construct an optimization dashboard enabling real-time visualization of both scalar and field quantities.

The article is structured into five sections. An introductory segment provides context for the problem and outlines pertinent theoretical foundations. Subsequently, the following section delineates the initial sizing of the air intake, the implementation of shape parameters, and ROM construction. The results section elucidates the selection of the optimal solution and its validation. Finally, a concluding segment summarizes the findings of the study and offers recommendations.

2 Theoretical Background

In this section, we offer a concise overview of the approach typically utilized in the preliminary phase of NACA air intake design. We then delve into the theoretical foundations of Radial Basis Functions and Proper Orthogonal Decomposition. These two mathematical frameworks serve as the foundation for both the shape parameterization method and the generation of ROMs.

2.1 Preliminary Design

The typical preliminary design approach of air intakes makes use of the ESDU86002 standard. It is based on semi-empirical formulations that allows to size and to estimate the performance of intakes for given input requirements and operative conditions. The main physical parameters involved are:

- Mass flow rate
- Momentum thickness
- Boundary layer thickness
- Mach number.

Figure 3 sketches the topology of a flush NACA intakes with a curved-divergent ramp. Its highly swept planform causes vortices to develop along the ramp sides introducing losses but generating a pressure recovery larger than the theoretical two-dimensional value. Its performance in general exceeds also the pressure recovery than can be obtained by other flush inlets. In [15] it is suggested a ramp angle (α) of 7° and an inlet width to depth ratio (w/d_t) of 4. w and d_t represent the width and height of the inlet section, as shown in Fig. 3. Such values usually allow to achieve better drag and pressure recovery performance. The other geometric values are to be evaluated based on the physical properties under examination.

The method is iterative, and the first step is to determine the inlet section. It starts with a first attempt value:

$$A_1 = \frac{\dot{m}}{\rho_0 V_0} \tag{1}$$

where $\dot{m}$ is the requested flow rate, while ρ_0 and V_0 are respectively the density and the velocity of the undisturbed fluid. A_1 is the initial value of the inlet area of the air intake.

Among the curves shown in the standard, only those associated with the range of interest were considered. The design point is the cruise point (M = 0.52). So, for example, in the diagram in Fig. 4, the points on the curves for M = 0.4 and M = 0.6 were evaluated, and the points for M = 0.52 were interpolate to derive a mathematical law.

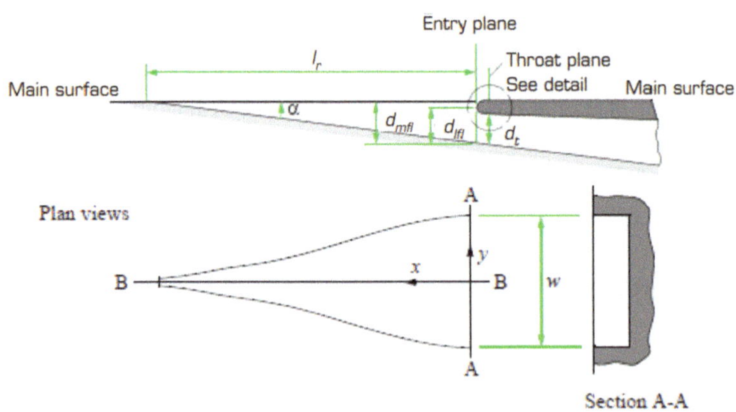

Fig. 3 Scheme of NACA air intake from ESDU86002

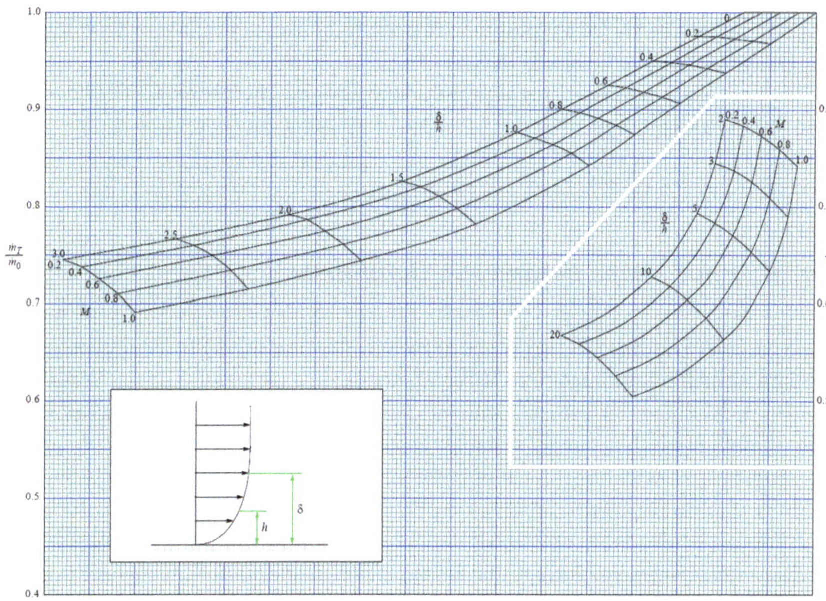

Fig. 4 Curves taken from ESDU standard

Here are the main steps followed for the sizing. For a more detailed explanation of all the steps, please refer to the reported standard [1].

Considering an elliptical lip:

$$d_t = \sqrt{\frac{0.25A_1}{1.125 - 0.25\tan\alpha}} \tag{2}$$

$$x = \frac{\delta}{dt} \rightarrow \frac{\dot{m}_t}{\dot{m}_0} \tag{3}$$

$$\dot{m}_i = \frac{\dot{m}_t}{\dot{m}_0} \cdot V_0 \cdot A_{1(i-1)} \tag{4}$$

$$A_{1i} = A_{1(i-1)} \cdot \frac{\dot{m}_{req}}{\dot{m}_i} \tag{5}$$

In these equations, the same nomenclature as the standard has been used. δ represents the boundary layer thickness and is an imposed boundary condition. In step (3), the diagram in Fig. 4 is used to calculate the ratio $\frac{\dot{m}_t}{\dot{m}_0}$, which represents the actual flow rate entering the air intake compared to the flow rate that could be processed.

These equations are iterated until A_1 and $\dot{m}$ converge. At the end of this process, section A_1 is obtained such that, considering losses and the shape of the air intake, the required flow rate is precisely processed. Once section A_1 have been identified from simple geometric considerations, the entire intake can be characterized. The final geometry obtained is given in Sect. 3.1.

This procedure was used for a preliminary design. The use of CFD allows for exploring a wider range of scenarios and identifying the optimum design more accurately [2–4].

2.2 RBF for Shape Parameters Definition

Shape parametrization [7, 8, 16] is one of the most critical aspect of a design problem based on numerical optimization. Its aim is to create a parametric model controlled by a limited number of values, reducing the variables space as much as possible while maintaining an accurate, reliable, and complete representation of the geometry. The proper selection of parameters strongly influences the result obtained. Incorrect choice of shape variations and a low-quality mesh updating procedure can compromise the entire optimization process, resulting in a non-optimal solution.

Mesh-morphing is a technique that allows acting on the numerical analysis domain changing the shape while keeping the topology of the grid unchanged. This is a critical aspect for ROM building. Radial basis functions (RBFs) find in mesh morphing their natural field of application, allowing the grid to deform efficiently and producing a final mesh of good quality. The shape modifications are defined by a system of radial basis functions which generates the nodal displacements from a set of source points. The source points are the locations at which the user prescribes the desired displacements. RBFs are used to interpolate the known displacements exploiting a proximity principle: the displacement at each grid node is assigned based on the distance from source points.

From a mathematical point of view [17], the RBF problem's solution consists of calculating the coefficients of a linear system, of order equal to the number of source points. These coefficients allow to define the displacement of an arbitrary node of the grid as the sum of the radial contributions of each source point. The result is an analytical solution in closed form that defines the displacement field at any point in the domain. This function will exactly have the values imposed at the source points and will interpolate those values at mesh nodes with a trend depending on the radial function used. Assuming m to be the number of source points, the interpolation function can be written as:

$$f(x) = \sum_{i=1}^{m} \gamma_i \phi(\|c_i - x\|) + p(x) \tag{6}$$

In Eq.6, the term $\phi(\|c_i - x\|)$ is the chosen radial function. So ϕ depends on the distance between the source point c_i and the node x to be interpolated. γ_i are the weight of the radial function, and p is a polynomial term. The polynomial p aims to regularize the solution. Its degree depends on the chosen radial basis. It can be shown that a unique interpolating function exists if the RBF is conditionally positive. If this condition is satisfied, and if the order is equal to or less than 2, the polynomial term is defined as follows:

$$p(x) = \beta_1 + \beta_2 x_1 + \cdots + \beta_{n+1} x_n \tag{7}$$

The unknowns of the system are the polynomial coefficients β_i and the weights γ_i, which can be computed by imposing the following conditions:

- The function has the known displacement value g_i at the source points:

$$f(c_i) = g_i \tag{8}$$

- The coefficients are orthogonal:

$$\sum_{i=1}^{m} \gamma_i q(c_i) = 0 \tag{9}$$

In matrix form, the problem can be written as:

$$\begin{bmatrix} C & P \\ P^T & 0 \end{bmatrix} \begin{Bmatrix} \gamma \\ \beta \end{Bmatrix} = \begin{Bmatrix} g \\ 0 \end{Bmatrix} \tag{10}$$

where C is the interpolating matrix containing all distances from RBF centres:

$$C = \phi(\|c_i - c_j\|), \quad 1 \le i \le m, 1 \le j \le m$$

P is the matrix $m \times (n+1)$ where the row j is defined as:

$$P_i = [1, x_1, x_2, \ldots, x_n].$$

For the 3D case

$$f^x(x) = \sum_{i=1}^{m} \gamma_i^x \phi(\|c_i - x\|) + \beta_1^x + \beta_2^x x_1 + \beta_3^x x_2 + \beta_4^x x_3$$

$$f^y(x) = \sum_{i=1}^{m} \gamma_i^y \phi(\|c_i - x\|) + \beta_1^y + \beta_2^y x_1 + \beta_3^y x_2 + \beta_4^y x_3$$

$$f^z(x) = \sum_{i=1}^{m} \gamma_i^z \phi(\|c_i - x\|) + \beta_1^z + \beta_2^z x_1 + \beta_3^z x_2 + \beta_4^z x_3 \tag{11}$$

with

$$x = \begin{Bmatrix} x_1 \\ x_2 \\ x_3 \end{Bmatrix}, c = \begin{Bmatrix} c_1 \\ c_2 \\ c_3 \end{Bmatrix}.$$

There is a wide range of available RBFs. The behaviour of the interpolation depends on the chosen RBF. Radial functions can be roughly classified in two families: the global supported and the compact supported ones. High accuracy is achieved using global support but involving higher computational costs. Compact supported RBF are used when long distance interaction is not desired. They can simplify the fitting process, as the matrix of the system becomes sparse, and can be managed with numerical algorithms specifically conceived for this class of problems [17].

2.3 POD and ROM Building

POD is a powerful technique widely used to extract important features from high-dimensional datasets. In the context of scientific applications, it has been successfully applied [9–14, 18, 19] to the reduction of computational expense, by establishing low-order models that accurately reproduce the behavior of the original system.

One of the most popular methods used in conjunction with POD [20, 21] is the singular value decomposition (SVD), which enables the compression of the dataset into a small number of modes. These modes correspond to the dominant features of the system's behavior, and are obtained by diagonalizing the covariance matrix of the dataset. M is the matrix composed of the learning data in which each column s_i represents the values of each individual snapshot, the SVD matrix can be written as:

$$M = U\Sigma V^* \tag{12}$$

Or in the extended form:

$$M = \overbrace{\begin{bmatrix} \vec{S}_0; & \vec{S}_1; & \ldots & \vec{S}_n \\ & & & \\ & & & \end{bmatrix}}^{\text{Number of learing snapshots}} \left.\right\} \begin{matrix}\text{Size} \\ \text{of field}\end{matrix} = \begin{bmatrix} \vec{U}_0; & \vec{U}_1; & \ldots & \vec{U}_n \\ & & & \\ & & & \end{bmatrix}$$

$$\times \begin{bmatrix} \sigma_0 & & 0 \\ & \ddots & \\ 0 & & \sigma_n \end{bmatrix} \times \begin{bmatrix} \vec{V}_0; & \vec{V}_1; & \ldots & \vec{V}_n \\ & & & \\ & & & \end{bmatrix}^* \tag{13}$$

where:

- Σ is a diagonal matrix composed of σ_i the singular values of matrix M.
- U and V are matrices such that: $U * U^* = Id$ and $V * V^* = Id$.
- U (respectively V) is composed of the left singular vectors U_i (respectively right singular vectors V_i) of M.
- s_i are the learning solution vectors (learning snapshots i).

The field solution X in the design space can then be approximated as a linear combination of r modes:

$$X = \sum_{i=1}^{r} \nu_i U_i \quad (14)$$

r being an important parameter that balances accuracy and compression without including noise. Two projection error curves, reducing error and leave-one-out (LOO) error, help determine an optimal number of modes. The first represents the mean error projection of the vectors used to calculate the r modes (SVD) (reduction error—blue curve in the following figure). The second one represents an average standard projection error provided by a "Leave One Out" process (LOO error—green curve in the following figure). Each learning vector is alternatively removed from the learning set, projected in the modes calculated with the remaining vectors, after which all these projection errors are averaged.

For small r values, these projection errors are almost the same. When r increases, the difference between the two errors increases (see Fig. 5). Based on this separation point, an optimal number of modes r is proposed.

Once the POD modes have been obtained, it is possible to construct a reduced order model (ROM), which is a low-dimensional mathematical representation of the system that preserves the essential dynamics of the original problem.

Overall, the POD-SVD method and ROM construction provide an efficient framework for the analysis of complex systems, enabling the extraction of essential features and the development of accurate, affordable models for predictive simulations. This part involves building a response surface of interpolation coefficients α_i where the Genetic Aggregation algorithm is employed [21, 22]. The Genetic Aggregation algorithm is an optimization method used to minimize error in evaluating response surface weights. Inspired by genetic algorithms, it iteratively refines parameters to enhance accuracy. Initially, a population of potential solutions is generated and evaluated based on fitness criteria. Through successive generations, individuals with higher fitness are selected, undergo crossover and mutation, producing offspring solutions. The process continues until convergence, yielding optimized weights for the response surface. This iterative approach efficiently navigates the solution space, improving the accuracy of weight evaluation for enhanced model performance.

The total error has two parts: reduction error, which can be enhanced by increasing modes and learning data set, and interpolation error, minimized by increasing the learning data set.

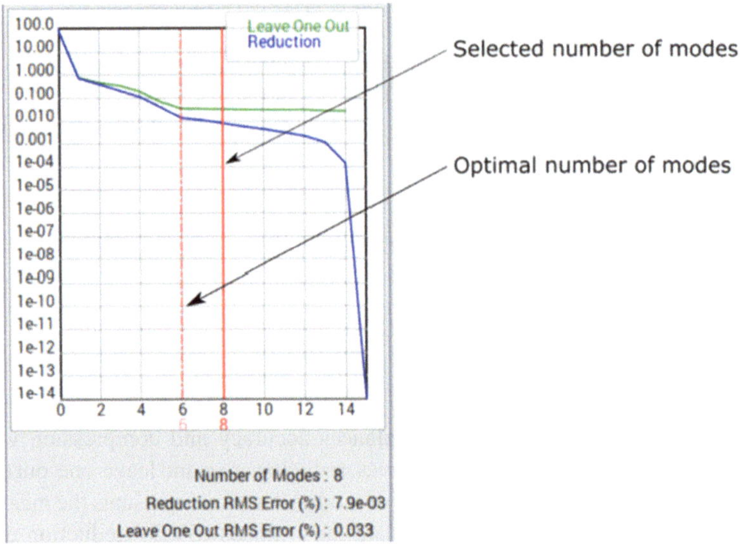

Fig. 5 Reduction and LOO error

3 Implementation of the Optimization Environment

The aim of this work is to create an optimization dashboard that allows real-time visualization of CFD results. Specifically, the optimization of an air intake is considered as a case study. The design point chosen is the cruise condition (M = 0.52), with the required mass flow rate for the cooling system known (0.23 kg/s). The idea is to start from the reference standard for a preliminary design that defines an already optimized geometry using a classical approach. Then, mesh morphing is used to refine the geometry and improve performance. The pressure at the outlet of the intake is chosen as the parameter to be optimized to minimize pressure losses and the work of any compressor downstream of the air intake. A critical aspect of this work was to develop a robust, generic, and easily replicable workflow for other case studies. The proposed workflow is illustrated in Fig. 6: a preliminary design was identified, on which CFD analysis was performed. Then geometric and physical parameters were defined and the ROM was created. Finally, the ROM was used to create the optimization platform.

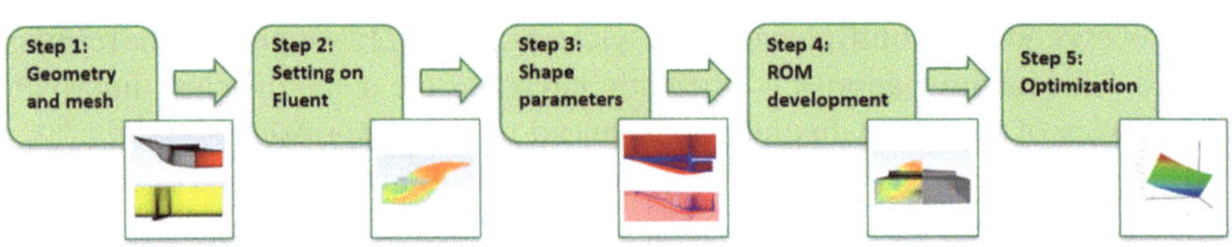

Fig. 6 Workflow

It can be noted that the main challenge lies in connecting and exchanging data between vastly different software tools and condensing inputs and outputs into a single dashboard. Another critical aspect is introducing shape parameters into a reduced model. In this regard, mesh morphing is the key, allowing for mesh deformation without altering its topology. Finally, using ROM for optimization enables a better understanding of the physics of the problem because real-time visualization is possible not only for scalar quantities but also for field quantities.

3.1 Geometry and Mesh

The ESDU86002 standard was used for the preliminary sizing of a starting reference NACA air intake according to defined operative working conditions. Such a geometry was used as baseline on which to generate the computational domain and to implement the shape parameterization. The intake was dimensioned to operate with a mass-flow rate of 0.23 kg/s at a Mach number of 0.52. The dimension of the generated starting duct is reported in Fig. 7.

3.1.1 Computational Domain

To reduce the size of the computational domain, we leveraged the geometrical symmetry property of the duct to generate a half-domain, assuming the development of a symmetric flow field. The NACA air intake was placed inside a volume 2 m long, 0.5 m wide and 0.5 m tall (Fig. 8). A multiblock structured grid composed of about 600.000 hexahedral cells was generated. Figure 9 displays the mesh on the symmetry plane and reports details of the cells clustering in the boundary layer and on the lip of the intake. The layering of the cells close to the wall was dimensioned assuming to adopt wall functions to model the boundary layer. In this way, it was possible to use a coarser mesh and reduce computational times.

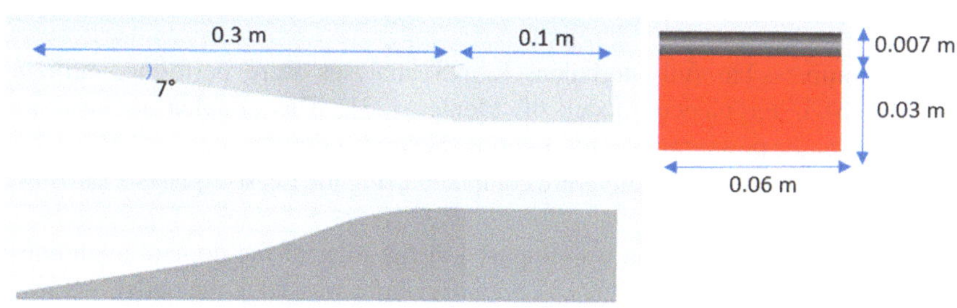

Fig. 7 Baseline NACA intakes geometry

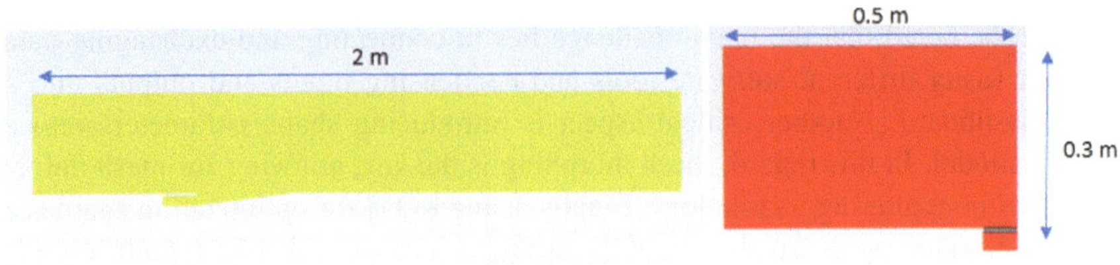

Fig. 8 Dimensions of the computational domain

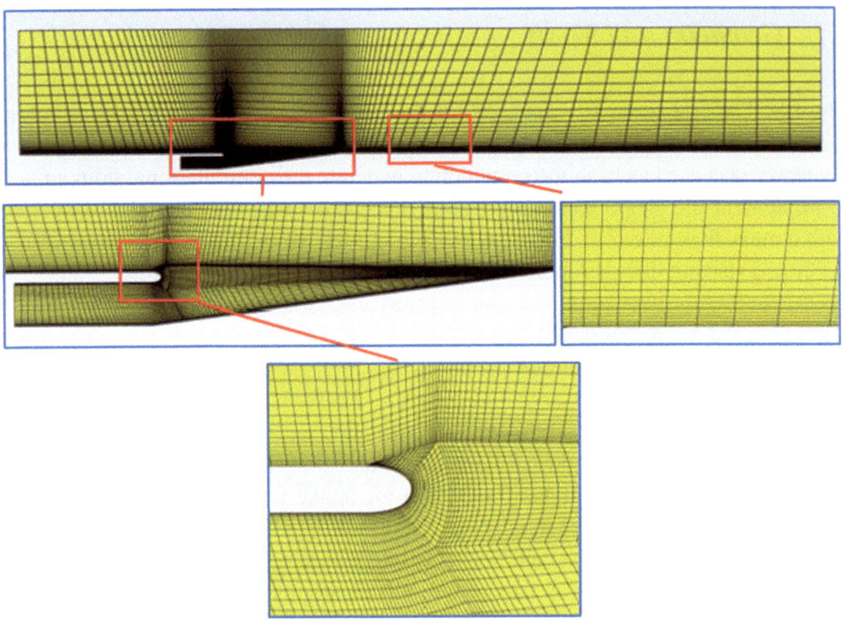

Fig. 9 Details of the computational grid

3.1.2 Boundary Condition

Steady compressible fully turbulent RANS analyses were performed utilizing the k-ω SST turbulence model of Menter [23]. The coupled scheme was employed for the pressure-velocity coupling. Second-order schemes were utilized for spatial discretization. Gradients were computed using the Least-Squares-Cell-Based method.

A *velocity inlet* condition was imposed at the inlet of the domain assuming a uniform velocity equal to 167 m/s. A *pressure outlet* was set at the outlet. All walls were modelled as smooth no-slip walls. Zero static gauge pressure was imposed at the outlet of the air-intake imposing the required target mass flow rate of 0.125 kg/s.

3.2 Shape Parameterization

The RBF Morph software [24] facilitates the intuitive definition of shape modifications by employing Radial Basis Functions (RBFs) to interpolate displacements specified on mesh nodes. Source points can be defined by directly selecting nodes on surfaces or by utilizing "encaps". Encaps refer to reference solids (spheres, cylinders, or parallelepipeds) within which points are positioned at user-defined uniform intervals. These encaps can serve to specify displacements (moving encaps) or delineate the region affected by the morphing process (domain encaps).

The *RBF Morph* software was used to define 8 shape parameters. The first 5 parameters control the ramp profile. Each of them acts locally. The overlap of the 5 parameters allows to change the shape of the ramp profile. A cylindrical moving encap was defined for each parameter, and a box encloses the fixed zone as shown in Figs. 10, 11, 12, 13 and 14. The preview of these 5 parameters is shown in Figs. 15, 16, 17, 18 and 19.

One parameter controls the ramp length and its angle. Again, a cylindrical moving encap was used to define the source points and displacements, fixed points were defined by directly selecting surfaces (Fig. 20). The preview is shown in Fig. 21.

One parameter controls the constricted area on the front plane. The same logic was used as for the previous parameters (Figs. 22 and 23).

Finally, the last parameter moves the entire lateral surface (Figs. 24 and 25).

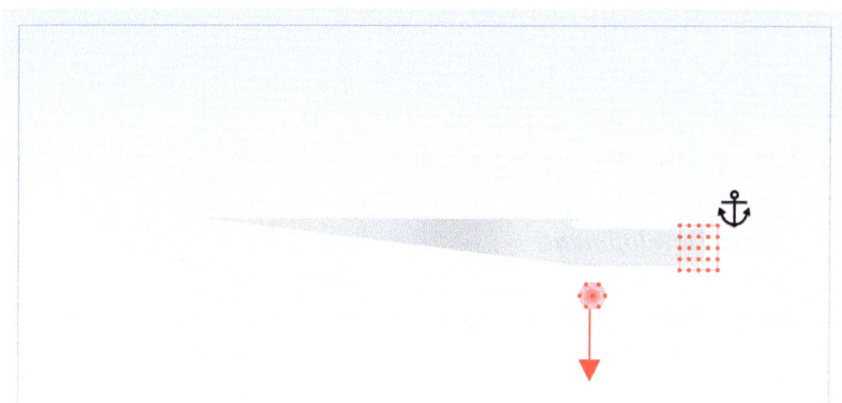

Fig. 10 Cyl_o1, setting. Blue encap and source points

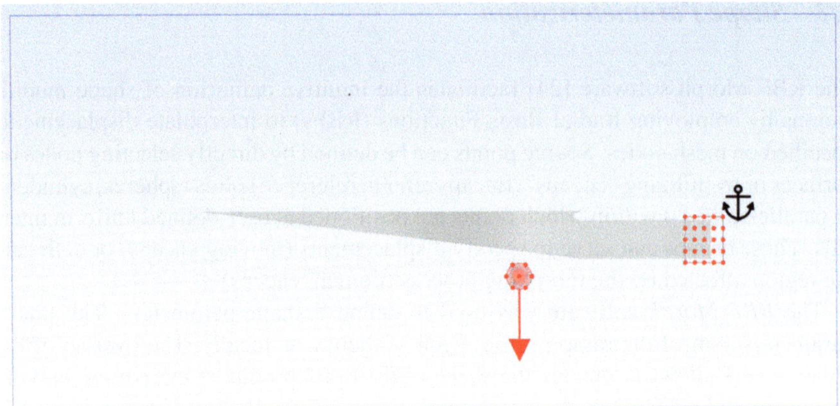

Fig. 11 Cyl_o2, setting. Blue encap and source points

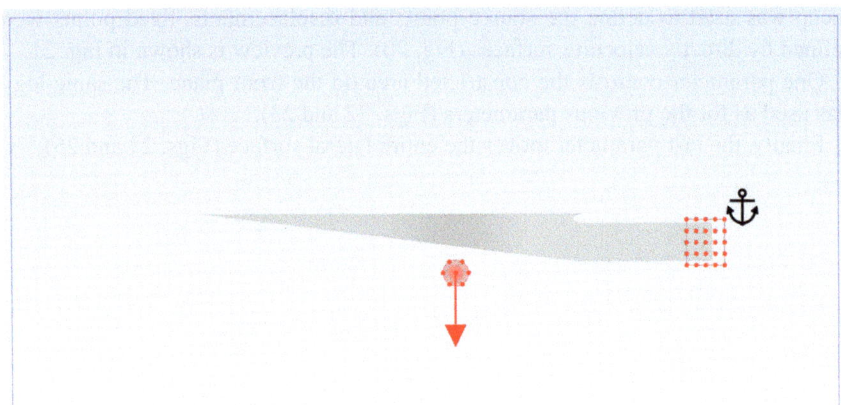

Fig. 12 Cyl_o3, setting. Blue encap and source points

3.3 ROM Development

The idea is to have an optimization platform where the best combination of shape parameters can be identified visualizing the flow field in real-time. To accomplish this, a ROM of the studied case was extracted. The following workflow was followed to create the ROM (Fig. 26):

DOE is the first step. For this study, 80 snapshots were used. The snapshots were identified using Latin Hypercube sampling [25]. Each snapshot comes from the solution of a CFD analysis of a geometry generated morphing the baseline mesh with a combination of shape parameter amplification factors representing a point of the variables design space.

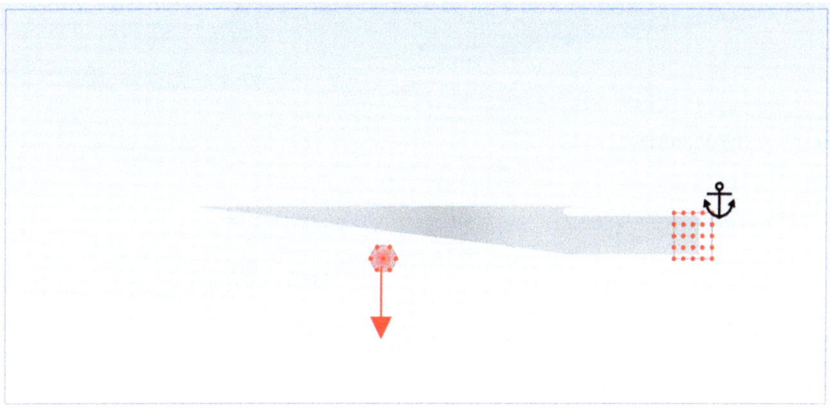

Fig. 13 Cyl_o4, setting. Blue encap and source points

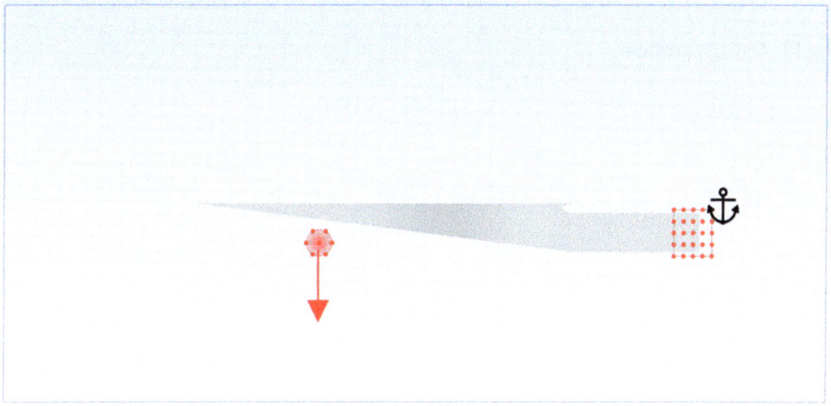

Fig. 14 Cyl_o5, setting. Blue encap and source points

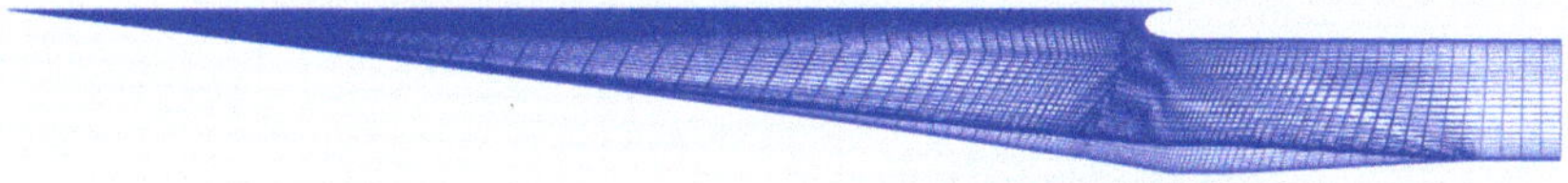

Fig. 15 Cyl_o1, preview

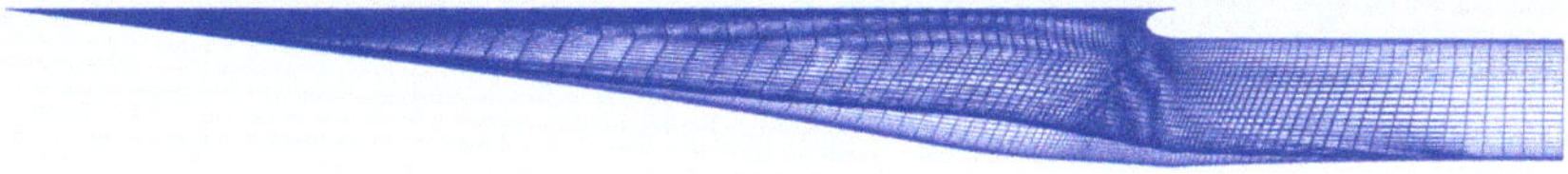

Fig. 16 Cyl_o2, preview

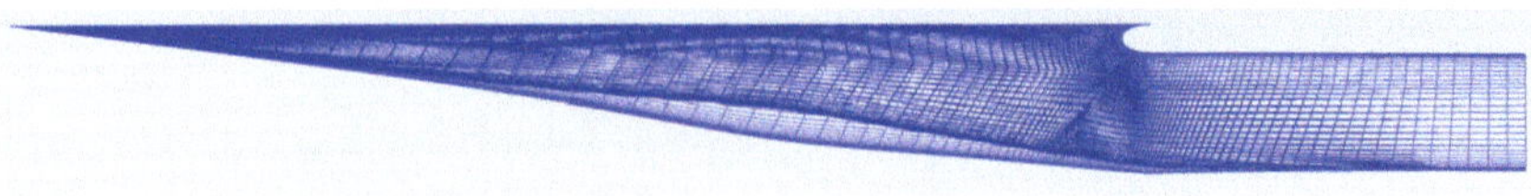

Fig. 17 Cyl_o3, preview

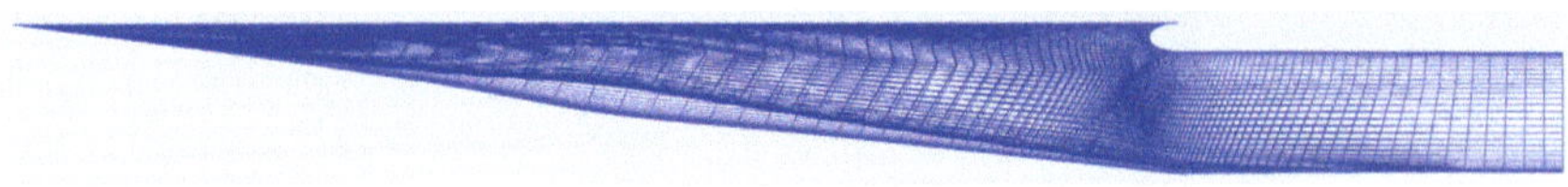

Fig. 18 Cyl_o4, preview

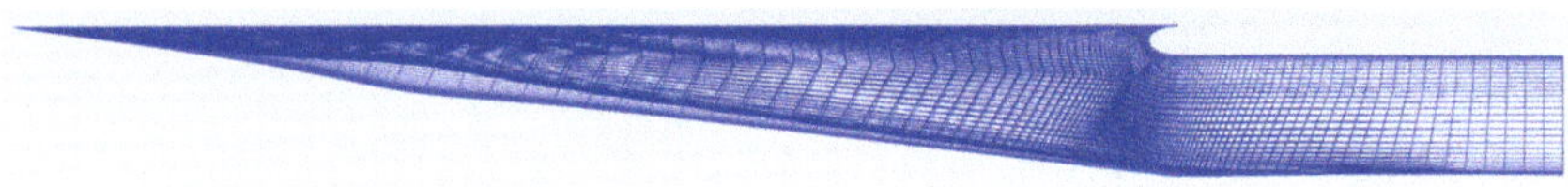

Fig. 19 Cyl_o5, preview

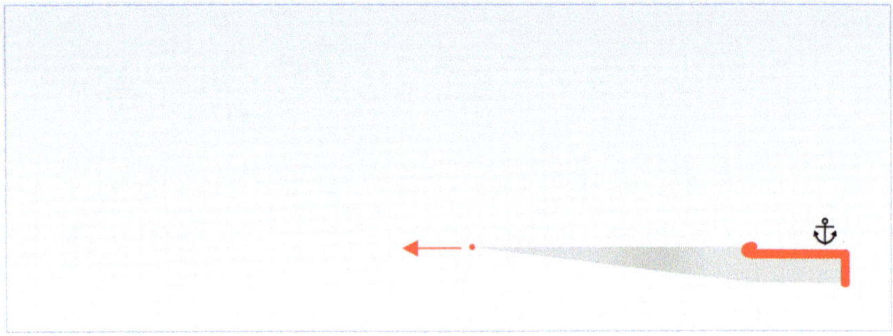

Fig. 20 Length, setting. Blue encap and source points

Fig. 21 Length, preview

The next step is data compression, aimed at reducing the number of variables in the system and simplifying the problem. Ansys Twin Builder [21] was utilized to perform POD and create the ROM. Ansys Twin Builder is software designed for creating static ROMs. In our case, each snapshot contains pressure values for every

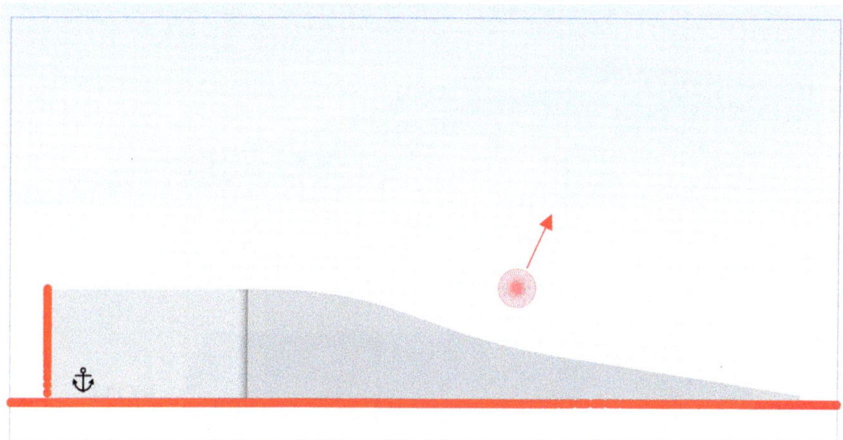

Fig. 22 Local_y, setting. Blue encap and source points

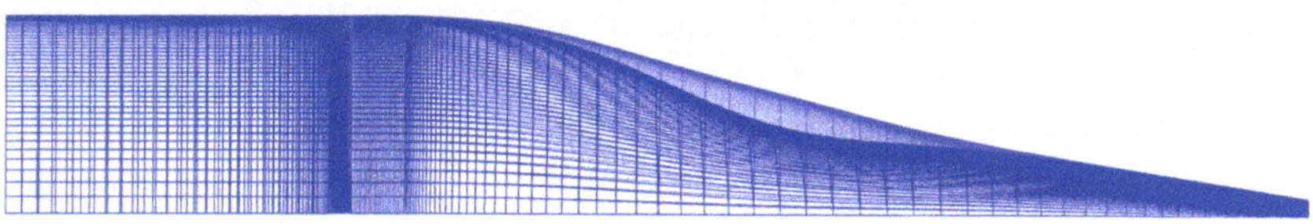

Fig. 23 Local_y, preview

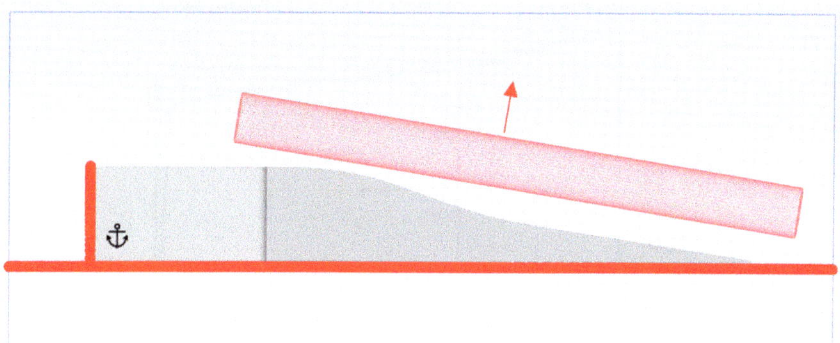

Fig. 24 Lateral, setting. Blue encap and source points

node on the air intake surface. The tool allows selection of the number of snapshots to be used for training, as well as the number of modes for reduction, while monitoring error at each step. The choice of these hyperparameters is then made based on a trade-off between error (considering underfitting and overfitting) and model complexity.

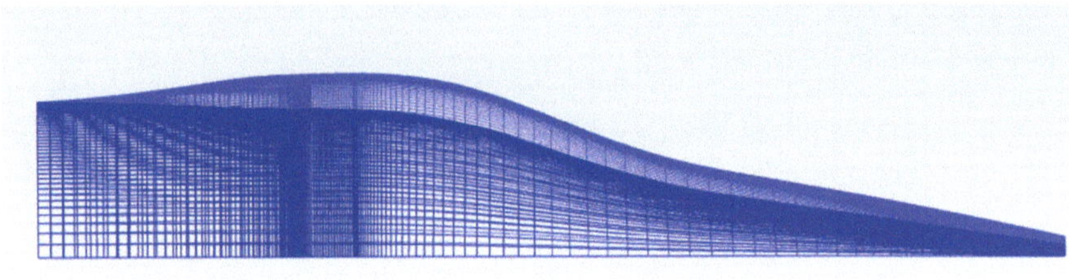

Fig. 25 Lateral, preview

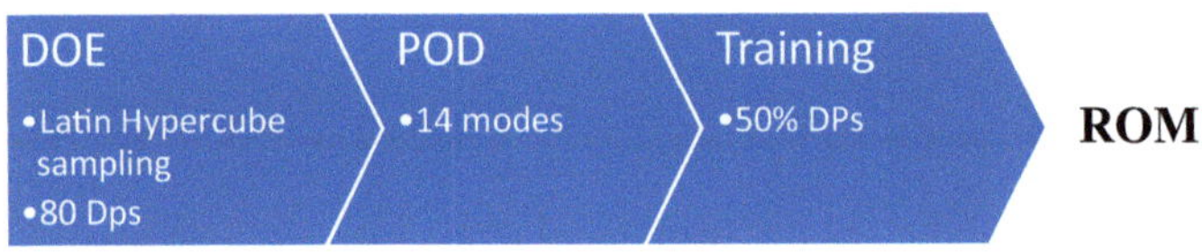

Fig. 26 ROM development workflow

So 14 modes were selected (Fig. 27) and the model was trained using 50% of the snapshots.

Once the ROM was created, it was possible to update the shape and the physical parameters. The solution field can be now visualized in real-time. Figure 28 reports the Ansys Twin Builder interface with which is possible to interactively update the

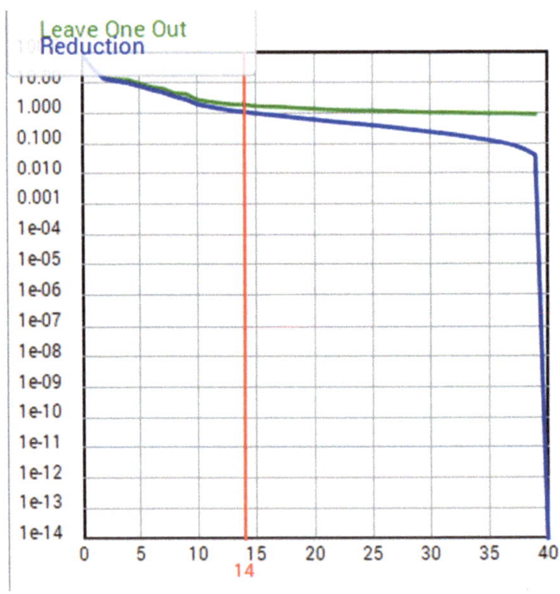

Fig. 27 Reduction and LOO error for the NACA ROM

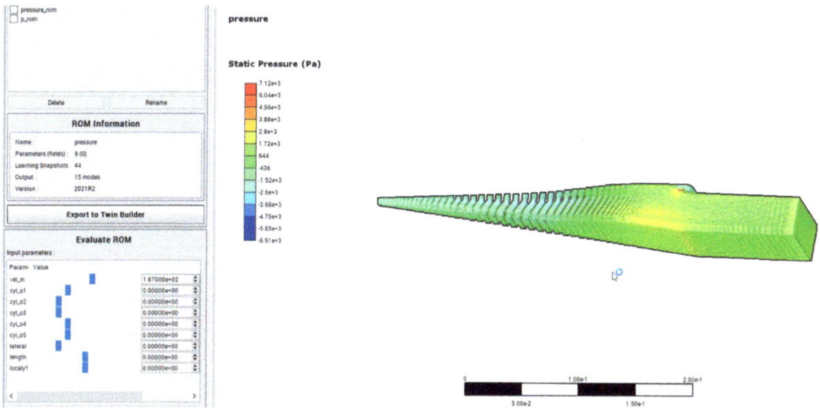

Fig. 28 Ansys Twin Builder interface

shape parameters amplification factors and to visualize in real time the updated geometry with its corresponding pressure field.

The obtained ROM was exported as a Functional Mock-up Unit (FMU) [26], imported into the Matlab-Simulink environment and coupled to an optimization algorithm. In this case, the gradient descent method [27] was used to identify the optimum. So, at each optimization step, a parameter variation of da = 0.1 is applied, and the ROM calculates the outlet pressure P_i. Then, the gradient is computed using finite differences: $g = (P_i - P_{i-1})/0.1$. A proportional amplification is then applied:

$$a_i = a_{i-1} + kg \tag{15}$$

Where k is a user-set multiplicative parameter. A larger k leads to faster convergence, but excessively large k values may result in poor sensitivity in detecting the maximum. This process is repeated for each parameter, and only after calculating the amplifications i for each parameter, the optimization proceeds to the next step.

An interface was created in Matlab environment (Fig. 29). The right part of the interface allows to maximize the pressure setting fixing the maximum number of iterations and the proportionality coefficient of the steepest descent method (k). During each iteration the values of parameters are updated in the proper boxes and pressure output calculated using the ROM. In the left part, it is possible to fix the parameter values manually. By the "*Evaluate*" button, the pressure field for the defined set of parameters is shown in real-time.

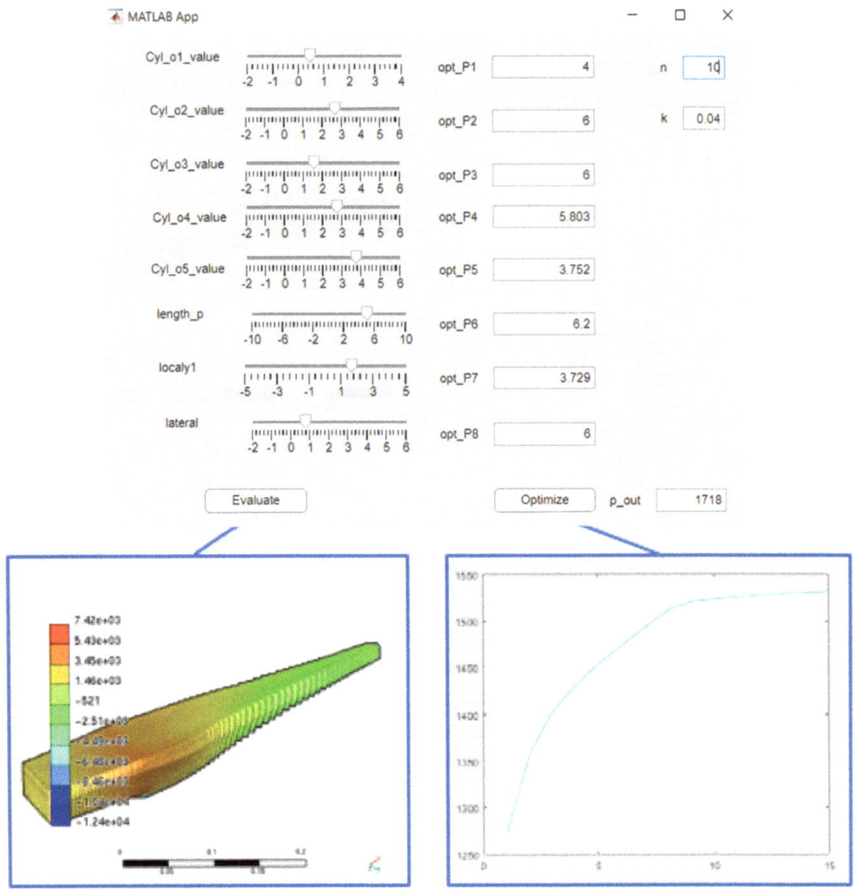

Fig. 29 Optimization dashboard

4 Results

To assess the quality of the proposed methodology, the results obtained with the use of ROM were compared to those achieved using plain CFD evaluations. In Fig. 30 the pressure contours obtained by the ROM are shown on the left, while the ones calculated using CFD are shown on the right for 3 points. Each point is a different combination of input parameters (the same value for CFD and ROM).

The total error has two parts: reduction error, which can be enhanced by increasing modes and learning data set, and interpolation error, minimized by increasing the learning data set. For each of the 80 DPs, the total error (for both the points used for training and the others) is calculated. The maximum error evaluated is in the order of 4% (Table 1).

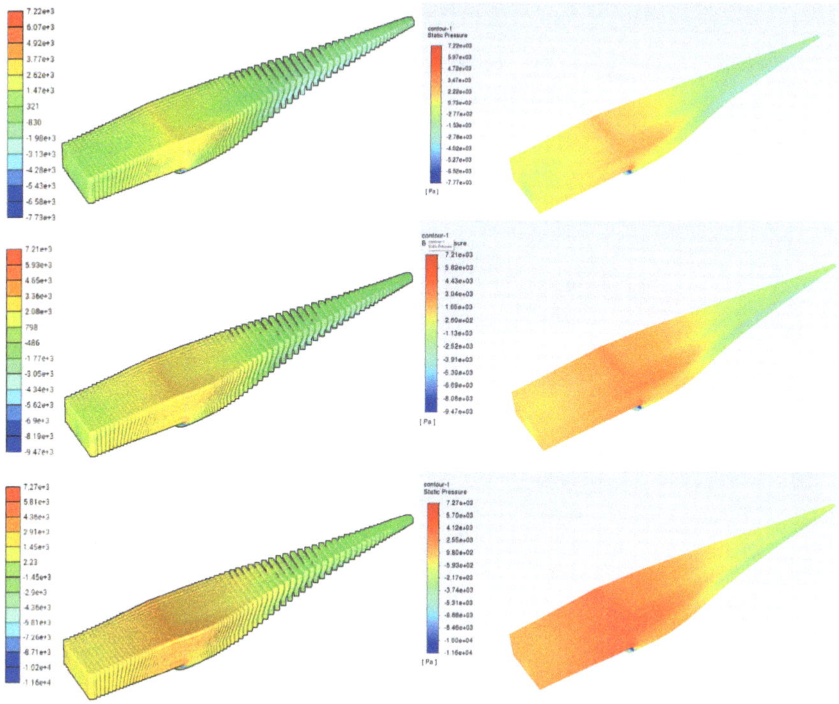

Fig. 30 Comparison between ROM (left) and CFD solutions (right) for 3 points

Table 1 Comparison of outlet pressure [Pa] between ROM and CFD for 3 points

ROM	CFD
722.2	743.1
1120.5	1165.3
1432.1	1482.2

Figure 31 compares the optimal design with the baseline geometry. Figure 32 compares the flow field and Figs. 33 and 34 displays the velocity and the pressure contours. The outlet pressure of the optimized geometry is 160% larger than the pressure on the outlet of the baseline geometry. Table 2 reports the averaged values of the pressure on the outlets. Finally, Table 3 reports the input parameter values corresponding to the optimized geometry and the ranges considered for each parameter. The range for each parameter was considered both by evaluating mesh quality (excessively large amplifications significantly reduce mesh quality) and by considering physical/topological assessments (excessively large amplifications may result in deformed final geometries with less meaningful shapes).

Looking at the reported results, it can be seen that in the optimized geometry, the flow is better guided at the duct inlet, resulting in a reduction in the maximum velocity

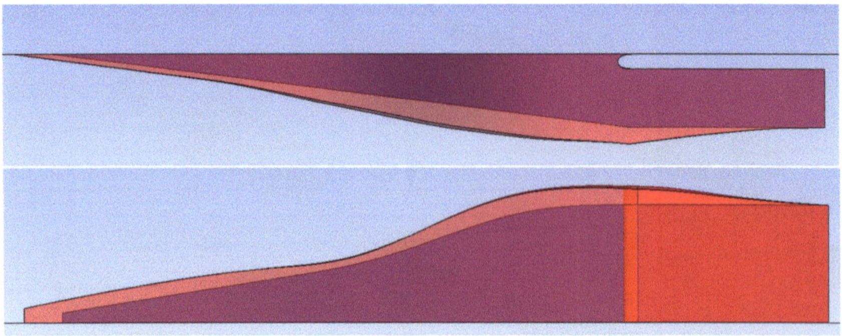

Fig. 31 Comparison of baseline (blue) and optimized shape (red)

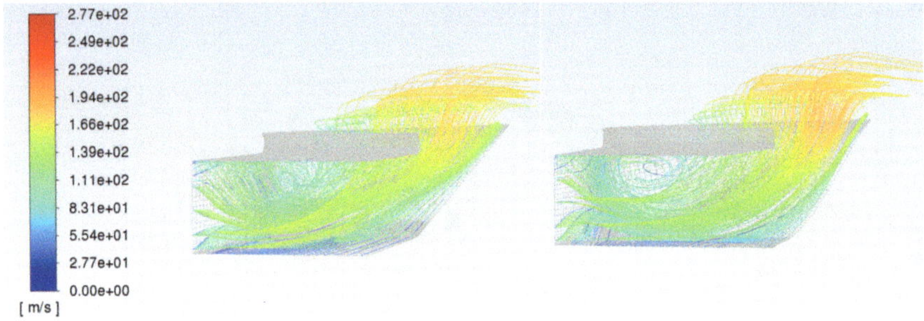

Fig. 32 Comparison of velocity streamlines of baseline (left) and optimized (right)

and separation recorded in the lip zone. In the baseline geometry, the flow tube has a strong bottleneck in the duct start zone that acts as a concentrated leak resulting in a significant pressure drop. This effect is limited in the optimized geometry.

5 Conclusions

This study explored the possibility of using ROMs to create optimization dashboards. Mesh morphing is a critical point as it enables the definition of shape parameters while preserving mesh topology. ROMs are highly versatile tools that can easily be integrated into any optimization workflow, allowing for real-time evaluation of both scalar and field quantities with high accuracy. Using this workflow, an optimization dashboard was created to make a change to the geometry and visualize the effects in terms of pressure distribution in real time. In addition, an optimal geometry is suggested using the ROM to evaluate the magnitude to be optimized and the gradient descent method to find the optimum. Similarly, the ROM of the velocity field could be constructed to have a greater understanding of the physics of the problem, and

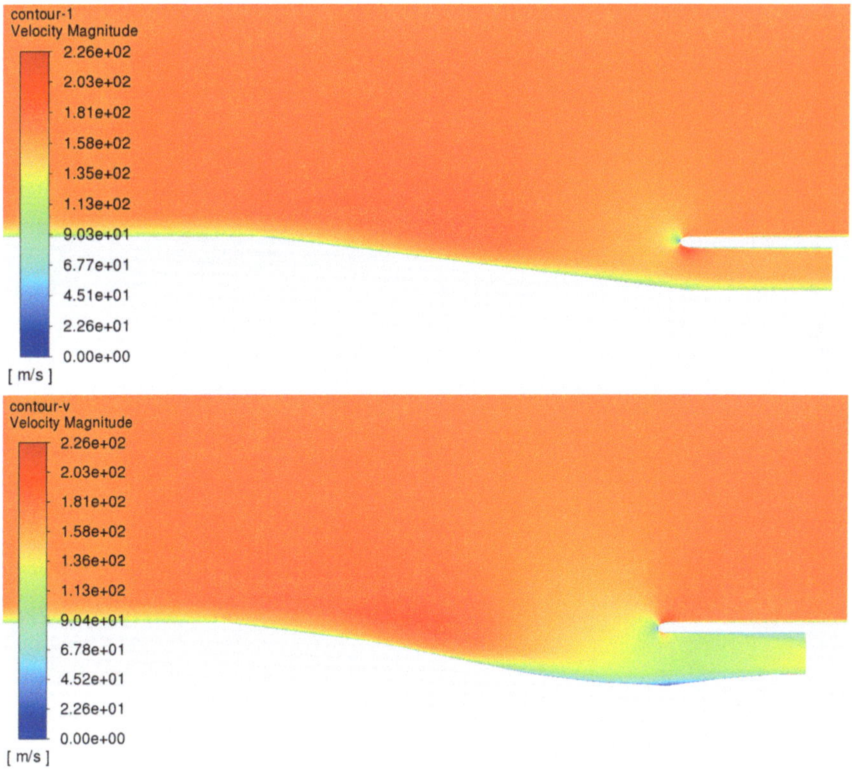

Fig. 33 Comparison of velocity contour of baseline (above) and optimized (below)

other scalar quantities, such as drag, could also be evaluated to implement multi-objective optimizations. Finally, the visualization of results could be enhanced. In this study, the Twin Builder viewer was directly used for visualization. As seen from the reported results, the tool employs a points-based graphics. A future development could involve utilizing a support mesh to enhance the visualization of pressure fields and integrating ROMs into virtual reality viewers for an even more immersive experience. Given that FMUs are the core of the optimization dashboard, being universal and easily transportable entities, this approach could enable the transport of CFD results augmented reality headsets, modify the mesh, and visualize the results in real-time within an even more interactive environment.

However, the results obtained were very satisfactory in terms of evaluation accuracy and versatility of the tool.

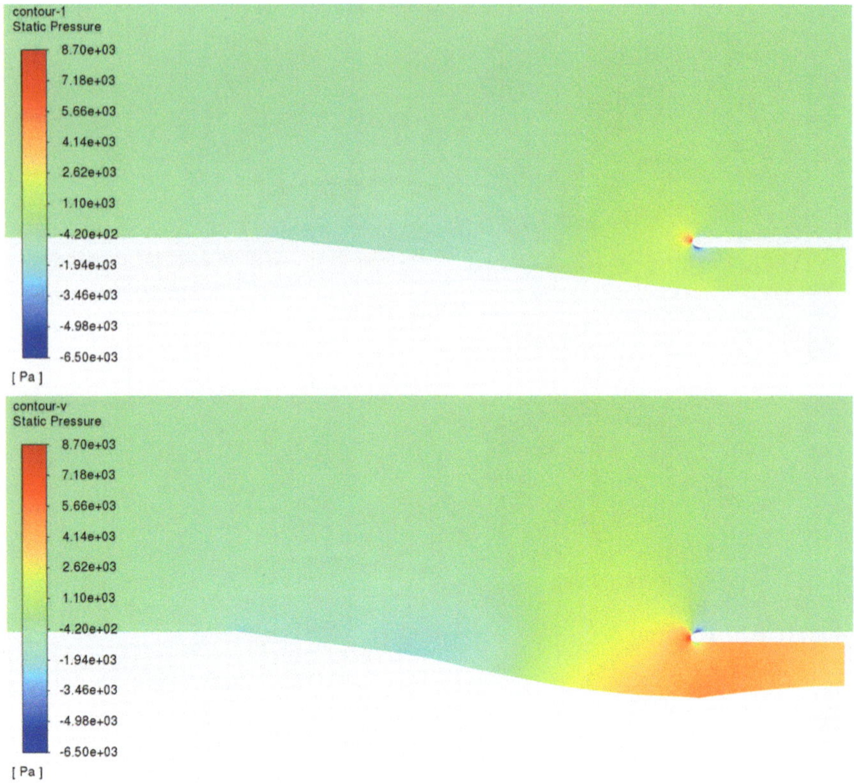

Fig. 34 Comparison of pressure contour of baseline (above) and optimized (below)

Table 2 Outlet pressure of baseline and optimized shape

	Baseline	Optimized
P_Out [pa]	664.5	1718.5

Table 3 Range and optimal value of each shape parameter

	Range	Optimum
P1	−2 to 4	4
P2	−2 to 6	6
P3	−2 to 6	6
P4	−2 to 6	5.8
P5	−2 to 6	3.8
P6	−5 to 5	3.7
P7	−10 to 10	6.2
P8	−2 to 6	6

References

1. ESDU 86002: Drag and pressure recovery characteristics of auxiliary air inlets at subsonic speeds (2004)
2. Da Silveira, B.H., Souza, P.R.C., Almeida, O.: Numerical investigation of a Naca air intake for a canard type aircraft. Int. J. Adv. Eng. Res. Sci. (IJAERS) (2017)
3. Zhu, D.X., Yang, Y., Yu, Z.J., Li, Z.P., Dong, V.: Optimization of the aerodynamic characteristics of a NACA air intake based on DoE and numerical methods. J. Phys. Conf. Ser. (2021)
4. Biancolini, M.E., Cella, U.: Flexible Engineering Toward Green Aircraft: CAE Tools for Sustainable Mobility (2020)
5. Viola, I.M., Biancolini, M.E., Sacher, M., Cella, U.: A CFD-based wing sail optimisation method coupled to a VPP. Edinburgh Research Explorer (2015)
6. Kapsoulis, D.H., Asouti, V.G., Papoutsis-Kiachagias, E.M., Giannakoglou, K.C., Porziani, S., Costa, E., Groth, C., Biancolini, M.E., Andrejašič, M., Eržen, D., Bernaschi, M., Sabellico, A., Urso, G.: Aircraft & car shape optimization on the RBF4AERO platform (2016)
7. Lopez, A.: Ottimizzazione di flussi esterni ed interni mediante metodi CFD adjoint e mesh morphing (2018)
8. Biancolini, M.E., Biancolini, C., Costa, E., Valentini, P.P., Gattamelata, D.: Industrial Application of the Meshless Morpher RBF Morph to a Motorbike Windshield Optimisation (2009)
9. Costa, A., Alba, M., Biancolini, M.E.: Geometric and physical reduced order modeling applied to CFD results (2022)
10. Biancolini, M.E., Capellini, K., Costa, E., Groth, C., Celi, S.: Fast interactive CFD evaluation of hemodynamics assisted by RBF mesh morphing and reduced order models: the case of aTAA modelling (2020)
11. Groth, C., Porziani, S., Biancolini, M., Costa, E., Celi, S., Capellini, K., Rochette, M., Morgenthaler, V.: The medical digital twin assisted by reduced order models and mesh morphing. In: International CAE Conference, vol. 10 (2018)
12. Ballarin, F., Faggiano, E., Ippolito, S., Manzoni, A., Quarteroni, A., Rozza, G., Scrofani, R.: Fast simulations of patient-specific haemodynamics of coronary artery bypass grafts based on a POD–Galerkin method and a vascular shape parametrization. J. Comput. Phys. **315**, 609–628 (2016)
13. Chaturantabut, S.A.: A state space error estimate for POD-DEIM. SIAM J. Numer. Anal. (2012)
14. Karpel, M.: Reduced-order models for integrated aeroservoelastic optimization. J. Aircr. (1999)
15. Rűtten, M., Krenkel, L., Freund, M.: Parametric Design, Comparison and Evaluation of Air Intake Types for Bleedless Aircraft. AIAA (2009)
16. Groth, C.: Adjoint-based shape optimization workflows using RBF (2017)
17. Biancolini, M.E.: Fast Radial Basis Functions for Engineering Applications. Springer (2017)
18. Gennaretti, M.A.: Study of reduced-order models for gust-response. J. Aircr. (2004)
19. Holmes, P.: Turbulence, Co-herent Structures, Dynamical Systems and Symmetry. Cambridge Monographs on Mechanics. Cambridge University Press (2012)
20. Weiss, J.: A tutorial on the proper orthogonal decomposition. In: AIAA Aviation 2019 Forum (2019)
21. Ansys: Twin Builder manual (2021)
22. Goldberg, D., Holland, J.: Genetic algorithms and machine learning (1988)
23. Menter, F.R.: Two-equation eddy-viscosity turbulence models for engineering applications. AIAA J. (1994)
24. RBF Morph Fluids User's Guide (2021)
25. Loh, W.L.: On Latin hypercube sampling. Ann. Stat. (1996)
26. http://fmi-standard.org/
27. Chong, E., Zak, S.H.: Gradient Methods, An Introduction to Optimization, 4th edn., pp. 131–160. Wiley, Hoboken. ISBN 978-1-118-27901-4 (2013)

Accurate Mapping in Fluid-Structure Interaction Analysis Across Models of Varying Fidelity Levels

Corrado Groth, Ubaldo Cella, Andrea Chiappa, Emiliano Costa, Giorgio Travostino, and Marco E. Biancolini

Abstract This paper addresses the aero-structural shape optimization of aircraft wings by integrating high-fidelity aerodynamic models that incorporate structural flexibility. We present a fluid-structure interaction (FSI) workflow designed for non-matching Computational Fluid Dynamics (CFD) and Computational Structural Mechanics (CSM) meshes, utilizing structural modes and Radial Basis Functions (RBF) for mesh morphing. The developed methodology is applied to an industrial case study: the winglet of the Piaggio Aerospace P180 Avanti EVO aircraft. The workflow successfully couples a hybrid Finite Element Analysis (FEA) model, which combines solid and lower-dimensional elements to represent the wing assembly (beam elements for the nacelle, shell elements for the wing box and solid elements for the winglet), with a high-fidelity CFD model.

Keywords FSI · Mode superposition · Mapping · Varying fidelity

1 Introduction

The aerospace industry faces a strong demand for faster design processes, precise product optimization, and reduced production costs. To meet these challenges, numerical simulations play a crucial role. Achieving realistic and high-fidelity results requires the simulation of all aspects of the real system, necessitating a multidisciplinary approach that incorporates the effects of all relevant disciplines in a single analysis. One significant category of these problems is Fluid-Structure Interaction (FSI), where deformable or movable structures interact with internal or external fluid

C. Groth (✉) · U. Cella · A. Chiappa · M. E. Biancolini
Enterprise Engineering Department, University of Rome "Tor Vergata", Rome, Italy
e-mail: Corrado.Groth@uniroma2.it

E. Costa
ENGYS, Trieste, Italy

G. Travostino
Flight Technologies Department, Piaggio Aerospace, Villanova d'Albenga, Italy

M. E. Biancolini et al. (eds.), *Digital Twins for Aerospace*, Springer Aerospace Technology, https://doi.org/10.1007/978-3-032-05022-9_7

flows [1]. FSI is particularly critical in wings design since even small deformations under aerodynamic forces can affect overall performance, thus making to consider fluid-induced deformations during design optimization essential.

Given the importance and complexity of FSI, many algorithms and methodologies were developed. Reviews of various FSI methods are available in the literature, including those by Morand and Ohayon [2], Dowell and Hall [3], Hou et al. [4], and Benra et al. [5]. FSI algorithms can generally be categorized into two main approaches based on the coupling degree between structural and fluid problems: the monolithic approach [6], which formulates and solves a single system of equations governing both structural deformations and fluid flow, and the partitioned approach [7], which uses two separate solvers with data exchanges between CFD and CSM models. In the partitioned approach, CFD and CSM meshes can be either matching or non-matching [4], necessitating load mapping [8–10] between fluid and structural domains. The monolithic approach is preferred for strong fluid-structure interactions [11], whereas the partitioned approach is suitable for weaker interactions [12].

For FSI problems involving small deformations, a simplified strategy called the mode-superposition approach can be used to avoid iterative data exchange between CFD and CSM solvers [13]. This method assumes that the structure deformation can be expressed as a linear combination of a limited number of structural modes, which are evaluated once using a CSM solver and then applied in all FSI cycles, thereby accelerating the process. The modal approach is particularly advantageous when CSM and CFD meshes do not match at the interface or are constructed using elements of different dimensionalities, such as beam elements for wing displacements. Mode-superposition allows for the application of deformation fields without requiring conforming elements to link the meshes degrees of freedom. The resulting deformation field, based on modal shapes, can be applied to the CFD mesh through surface and volume morphing using RBF, a proven fast and reliable method for mesh morphing across various applications [14–16].

The framework employed in this study leverages well established mesh morphing techniques for aerospace and other applications [17–21]. Specifically, the FSI mode-superposition method, whose efficiency was widely demonstrated in facing several transient applications [22, 23] and static problems [24, 25], is used to address the multi-physics nature of the problem. This approach, initially verified on a simple, bulk wooden propeller blade [26], is extended here to aircraft wings, incorporating a complex internal wing box assembly represented by a simplified numerical mesh that does not match the corresponding one of the high-fidelity CFD model. In such cases, mismatches between structural and fluid interfaces can arise due to differing mesh spacings, levels of detail, or missing parts for structural reasons, thus necessitating more sophisticated load mapping procedures [27]

Furthermore, the proposed workflow is designed to be solver-agnostic, that means capable of integrating various tools including both open-source and commercial software. The paper is structured as follows: Sect. 2 outlines the theory behind both mode superposition and RBF required for mesh morphing, Sect. 3 details the integration of mesh morphing with modal theory to develop an aeroelastic numerical model, and Sect. 4 presents a comprehensive industrial application of the proposed method.

2 Theoretical Background

2.1 Mode Superposition Method

The mode-superposition approach is a well-known and powerful theory in structural engineering which a wide range of dynamic analyses are based upon. Both free and forced vibration problems can be solved using this method, without limitation on the system type (continuum or discrete) to be solved. According to this theory, undamped free vibration modes can be used to uncouple the equations governing the vibrational response. Each vibration mode is characterised by a natural frequency (eigenvalue) and a displacement field (eigenvector), which can be obtained through a modal analysis. For a given number n of modes, the following listed four steps are required to be carried out in sequence:

1. n eigenvalues and eigenvectors are computed on the wetted surfaces at the cost of a single linear analysis using the CSM solver;
2. aerodynamic loads are computed over the wetted surfaces using the CFD tool;
3. for each one of the n so-computed modes, the response under aerodynamic load is evaluated at the cost of n products;
4. a limited set of modal responses, ordered according to increasing eigenvalue value, is chosen and responses are superposed to determine the full modal response to the aerodynamic load.

The governing equation of a structure seen as a discrete system with n degrees of freedom (DoFs) and without accounting for damping can be written as:

$$M\ddot{x} + Kx = f(t) \tag{1}$$

where x is the displacement vector, $\ddot{x}$ is the acceleration, M is the mass matrix of the system, K is the stiffness matrix and $f(t)$ is the time-varying loading vector. The mass and stiffness matrices are the discrete approximation of the distribution of mass and local stiffness of the system, computed adopting the Finite Element Method (FEM). Assuming a harmonic solution $x = ue^{i\omega t}$ as a possible response of the system for each DoF, Eq. (1) can be rewritten as [28]:

$$Ku = \omega^2 Mu \tag{2}$$

that represents the free vibration of a system after loading, ω being a set of eigenfrequencies, one for each DoF. Once Eq. (2) has been solved, it is possible to find a complete set of eigenvectors u and eigenvalues ω^2 for each DoF which can be converted into frequencies ν, since $\omega = 2\pi\nu$.

Furthermore eigenvectors (i.e. modal shapes) represent an orthogonal basis of the eigenvalue problem and the dynamic response of the system can be represented as [29]

$$x = U\eta \tag{3}$$

in which U is the $n \times q$ matrix composed by the q modal shapes u and η is a $q \times 1$ vector of modal coordinates, that represent the weights used in the sum of all modal shapes in the mode-superposition approach. A convenient way to manage the orthogonalisation of the governing equations is through the normalisation of the modal mass with respect to the orthogonal basis. For a given mode m it is:

$$u_m^T M u_m = 1, \qquad m = 1, 2, \ldots, n. \tag{4}$$

that leads to n decoupled single DoF system equations [30]:

$$\ddot{\eta}_m + \omega_m^2 \eta_m = u_m^T f(t) \tag{5}$$

that can be used to find the needed modal coordinates η_m. Each nodal force defined as

$$N_m(t) = u_m^T f(t) \tag{6}$$

can be obtained by multiplying mode eigenvector u_m by the real physical force distribution $f(t)$. At the end, the complete response of the system to a given load can be obtained using Eq. (3), since all entities are known.

Dealing with continuum mechanics, the number of eigenvalues and eigenvectors are infinite. In practice it is not necessary to use all mode shapes to describe the deformation of a given body, since the eigenvectors associated to the higher eigenvalues have a lower energy and contribute less to the total deformation. Hence, a good approximation in system deformation can be achieved with only a few first most energetic mode shapes, which tend to provide great contribution to the structural response [29].

The mode superposition approach is principally used in dynamic analyses; when dealing with static loads, the solution can be approximated by simplifying Eq. (5) into the form:

$$\omega_m^2 \eta_m = N_m. \tag{7}$$

In Sect. 3 it will be demonstrated how such a modal approach is a key enabler of the FSI-based optimisation proposed in this work when employed synergistically with an RBF mesh morpher. Equation (7) can be adopted to compute the intensity of modal coordinates η_m in response to modal forces N_m due to the applied aerodynamic loads. Once η_m are available, the deformed shape is retrieved according to Eq. (3).

2.2 Radial Basis Functions

In this work, the strategy chosen to deform the CFD mesh according to the changed shapes is based on RBF mesh morphing. Mesh deformation is indeed required in order to modify the numerical domain according to the structural modal shapes for the coupled FSI approach, as described in Sect. 3.4, and according to the shape modifications. The RBF mathematical tool is used to interpolate the displacements defined at a set of discrete points (source points). RBF interpolation ensures a smooth morphing of the CFD mesh while applying the exact displacement of the source points. Once source points are defined either on the surfaces or in the volume, and boundary displacements are set, the RBF problem, written as a linear system with dimension equal to the number of source points, is solved. Once the system coefficients determined, it is possible to compute the displacement of each node as the sum of radial contributions of all source points. Therefore, the position of each node of the computational mesh is modified while preserving the computational grid topology. The interpolation function is composed by an RBF and a polynomial H of order $h-1$, where h is the order of the RBF, needed to preserve compatibility with rigid motion and guarantee the uniqueness of the solution; this polynomial is written as:

$$I(r) = \sum_{i=1}^{S} \gamma_i \varphi\left(\left\|r - r_{k_i}\right\|\right) + H(r) \tag{8}$$

where r is the position vector of any mesh node, r_{k_i} is the i-th source point position vector (S is the total number of source points) and $\|\cdot\|$ is the Euclidean norm. It has been demonstrated that a unique interpolant exists if the RBF is conditionally positive definite [31]. In this case, the linear polynomial

$$H(r) = \beta_1 + \beta_2 x + \beta_3 y + \beta_3 z \tag{9}$$

is used, for $h = 2$ [32]. So, it is possible to find an RBF fit if the RBF coefficients γ_i and the weights β_i of the polynomial terms are computed so that the imposed displacements g_i are reproduced at source points and the orthogonality conditions satisfied:

$$I\left(r_{k_i}\right) = g_i, \quad 1 \le i \le S; \qquad \sum_{i=1}^{S} \gamma_i q\left(r_{k_i}\right) = 0 \tag{10}$$

for all polynomials q with a degree less or equal to that of polynomial H. Coefficients γ_i and β_1 are obtained by solving the linear system

$$\begin{pmatrix} R & P \\ P^T & 0 \end{pmatrix} \begin{pmatrix} \gamma \\ \beta \end{pmatrix} = \begin{pmatrix} g \\ 0 \end{pmatrix}, \tag{11}$$

in which R is the interpolation matrix defined by the radial interactions between source points

$$R_{ij} = \varphi\left(\left| r_{k_i} - r_{k_j} \right|\right), \quad 1 \le i \le S, \quad 1 \le j \le S \tag{12}$$

and P is a constraint matrix which arises from orthogonality conditions

$$P = \begin{pmatrix} 1 & x_{k_1} & y_{k_1} & z_{k_1} \\ 1 & x_{k_2} & y_{k_2} & z_{k_2} \\ \vdots & \vdots & \vdots & \vdots \\ 1 & x_{k_N} & y_{k_N} & z_{k_N} \end{pmatrix}. \tag{13}$$

RBF based interpolation is suitable for scalar fields. In order to interpolate a displacement in three directions, the linear system of Eq. (11) is solved three times, separately for each displacement field. The mesh morphing approach based on RBF is adopted in this work as a part of the FSI coupled solver, as it allows to embed the structural modes into the CFD solution process.

3 High Fidelity Multi-physics Solvers

The coupled FSI approach addresses the aeroelastic problem in a compact and embedded manner. This section first outlines the methods used to solve the fluid flow and structural problems individually, followed by an explanation of how the mode superposition technique integrates them into a unified FSI solver. RBF mesh morphing serves as the crucial link between the CFD and CSM environments, and the subsequent sections focuses on the challenges encountered during its implementation across different multi-physics solvers. A key feature of the proposed workflow is its flexibility regarding software integration, allowing the incorporation of any commercial, open-source, or in-house CFD solver. This flexibility is facilitated by the meshless nature of RBF, which simplifies the coupling process.

3.1 CFD Solver

The core concept of the proposed workflow is to leverage the mode superposition method, thus making the CFD domain structurally flexible by embedding structural modal shapes. This is achieved by updating the CFD mesh through the computation of modal forces, as outlined in Eq. 6. The resulting modal coordinates are used to scale each mode, deforming the computational domain via the RBF method, as detailed in Sect. 2.2, before proceeding with the simulation. This approach effectively creates an embedded FSI solver where fluid dynamics and structural responses are evaluated in sequence. Any CFD solver that accurately captures the relevant physics—whether in-house, commercial, or open source—can be integrated into this framework. However, it is crucial that the solver allows sufficient customization for the user to compute modal forces and apply the resulting deformations to the mesh directly.

In the application presented in this study the CFD code used was ANSYS® Fluent®. The interaction with the solver was implemented through User Defined Functions (UDFs) written in C. The simulation was controlled via scripts written in Scheme language. The proposed workflow is versatile and has been successfully implemented by the authors using other CFD solvers, including HELYX, CFD++ [33], OpenFoam [34] and SU^2 [13] among others.

In cases where deep solver interaction is not feasible, a less efficient implementation can be employed by computing modal forces and deforming the mesh outside the CFD solver environment. This approach typically involves transferring data through external ASCII files, where load distributions are exported at specific spatial points for processing the modal force calculations. This method is less efficient due to the extensive I/O operations, the need to stop and restart the solver for each FSI update, and the requirement to perform new RBF interpolations using Eq. 8 at each FSI cycle. This strategy is often necessary when using commercial "black-box" software that restricts data exchange with the CFD solver.

3.2 CSM Solver

In the proposed workflow, which utilizes the mode superposition method, the structural solver is employed to extract modal shapes and frequencies. Since any structural displacement can be expressed as a linear combination of its eigenvectors, the modal analysis must be performed under boundary conditions consistent with those handled during the aeroelastic simulation. Any CSM software capable of exporting the positions of wetted surface nodes and modal results can be utilized.

A crucial aspect of implementing a mode-based FSI analysis is the verification of the modal basis, addressing one of the unresolved issues from the NASA Aeroelastic Prediction Workshop (AePW) [35], which concerns determining the optimal number of modes for static analysis. For example, DLR used 20 modes for its aeroelastic analysis at the workshop [35] but later increased this number to 30. However, a general guideline for the optimal number of modes cannot be prescribed, as it depends on the specific case and requirements, including the range of excited frequencies and load conditions. Some modes may exhibit local deformations that are not excited by the FSI loads, or symmetrical deformations in the case of a half-symmetric CFD problem. For transient FSI cases, attention must also be paid to the range of frequencies relative to the expected loads. To address this, the authors proposed a modal basis qualification procedure in [25] to identify the minimal number of modes necessary to achieve accurate deflections and aerodynamic coefficients.

In typical aeronautical applications, another challenge is the differing levels of detail required by structural and fluid dynamic analyses. In the aerospace field, structural models with varying levels of detail are often employed, featuring simplified representations in some areas and more detailed ones in other areas. In certain regions, the structural model could be simplified at such extent that it does not overlap with the wetted surfaces used in CFD analysis. This discrepancy poses significant challenges when using classical two-way FSI methods, where aerodynamic loads need to be transferred onto the wing structure such as ribs and spars, and structural displacements need to be synchronized back to the CFD model. This process can be highly complex or even impractical. In contrast, as the mode superposition method requires only a single structural computation before the FSI coupling and not necessitates continuous data transfer between the fluid dynamics and structural simulations, it can be more easily used in cases where the geometries of the fluid and structural simulations do not overlap. The problem of transferring the structural modes obtained by means of a simplified model to the CFD remains.

The proposed workflow overcomes this issue by using RBF as a morphing tool to interpolate the CSM modal displacements of the simplified structure onto the fully detailed boundaries of the CFD model and propagate these deformations over the fluid volume domain. Modal shapes are exported from the structural model of the CSM solver even if the nodes do not overlap with fluid boundaries, as illustrated in Fig. 1 for the RIBES wing [36], where nodes extracted only on the spars and ribs (left) are superposed onto the CFD geometry in grey. Using the skin, as shown on the same figure (right), would have been a better option, but not always available depending on the level of detail of the structural model itself.

3.3 RBF Solver

As detailed in Sect. 2.2, RBF have several attributes that make them highly suitable for mesh morphing applications. Being a meshless method applied only to mesh nodes, RBFs are not constrained by the element types used and are well-suited for parallel implementation [37]. Once the RBF problem is solved, each CPU core can independently apply mesh node displacements without the need of exchanging information from other partitions, as RBFs inherently ensure continuity at interfaces. This method allows precise control over geometry deformation by specifying the desired displacement directly at source points on the geometry surfaces. Users can impose a prescribed displacement or leave regions undeformed simply by setting a zero displacement. For mesh morphing, the equations system (11) is solved separately for each displacement component of interest. Nodes exported from the CSM solver, as described earlier, serve as source points to create an RBF problem for each modal shape, which is then applied to update the CFD mesh geometry under loads.

Nevertheless, using only the nodes from the wetted surfaces exported by the CSM solver is insufficient, as a more complex setup, including constraints on fixed surfaces, is often necessary. RBF-based deformation propagation within volume elements can be controlled by configuring the RBF problem appropriately, such

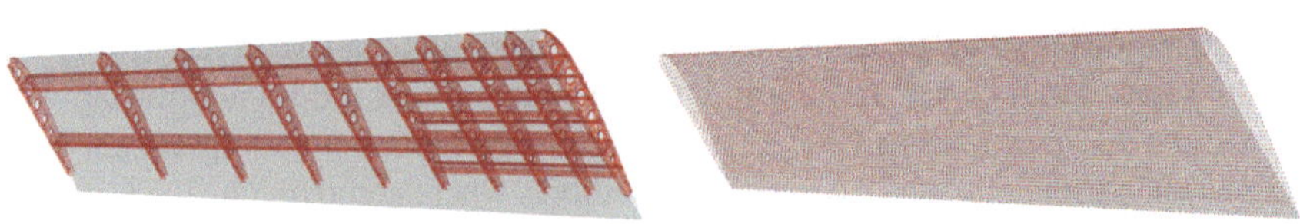

Fig. 1 Left: source points using the wing box, right: source points on the wetted surface

to allow morphing to affect only specific portions of the volume mesh while leaving other areas undeformed. The approach suggested here is similar to those described in [13] and [38], where the RBF setup is divided into fixed and moving source points. The moving set uses node positions and displacements extracted from the CSM solver only where needed, while the fixed set includes stationary points that constrain the deformation of fixed surfaces. Source points can be applied to surfaces or portions of geometries that need to remain undeformed, effectively constraining their motion. Volume smoothing is managed by defining fixed points within the volume mesh, limiting the morphing effect to targeted regions. A free deforming buffer area is set to accommodate deformations between moving and fixed surfaces.

Figure 2 illustrates an example of the RBF setup with fixed and moving sets for the first modal shape of the HirenaSD case [13] from the NASA Aeroelastic Prediction Workshop. The fixed set (left) constrains the fuselage surfaces and prescribes zero motion at a defined distance from the fuselage and wing. The moving set (middle) includes source points exported from the CSM solver. In this example, the motion applied to the green source points on the wing corresponds to the first modal shape (blue points strongly amplified for clarity). The complete setup, which blends both fixed and moving sets, is shown on the right of Fig. 2. It features a buffer zone near the wing-fuselage junction, thus allowing relative displacement of the wing root, which was attached to a damper in the structural model.

3.4 Coupled FSI Solver

The FSI procedure outlined in this paper consists of three stages, as introduced in Sect. 2.1:

1. Calculation of vibration modes;
2. RBF-based morphing computation for each vibration mode;
3. iterative FSI cycles involving a CFD solver and RBF-based mesh morphing.

The first two stages serve as preprocessing steps, completed once before the FSI loop begins. As depicted in Fig. 3, the initial stage uses the undeformed geometry to calculate the structural vibration modes via a structural solver. These modes are then utilized in the mode superposition phase. During this stage, the number of modes

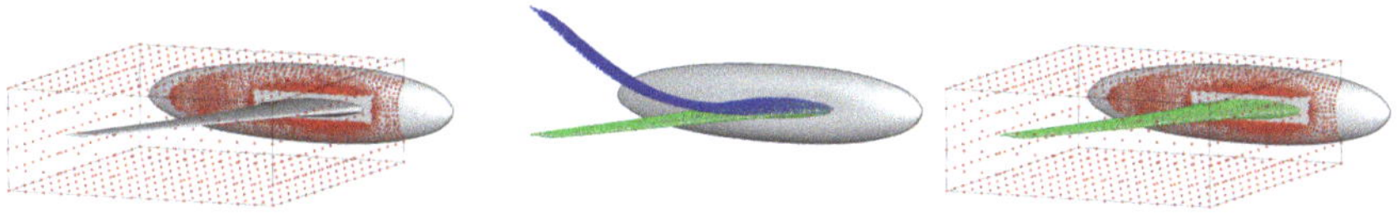

Fig. 2 Morphing setup using fixed and moving sets. Left: fixed set, center: moving set with original (green) and final positions (purple), right: RBF points distribution resulting from the addition of fixed and moving sets

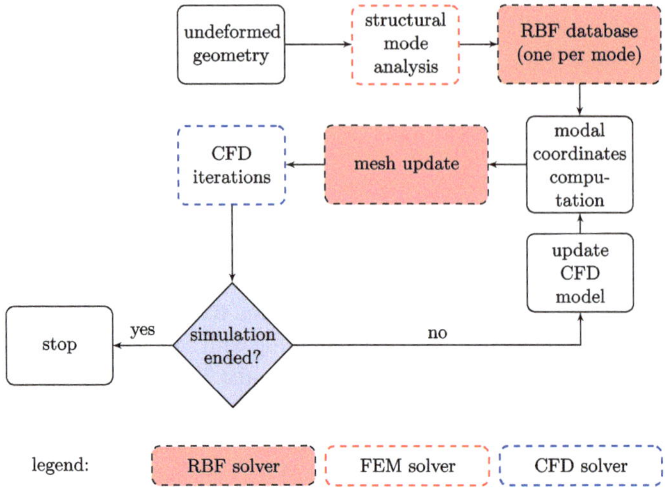

Fig. 3 Proposed FSI workflow based on embedded structural modes

to include is selected based on their energy contribution and impact on the overall structural displacement.

In the second stage, an RBF solution is computed for each vibration mode using the modal displacements as inputs. FEM nodes from the CSM model act as source points, and modal shapes are used as the displacement field. Zero displacement is assigned to the nodes corresponding to the rigid parts of the structure. Careful consideration must be given to the density and placement of source points, as they can impact the quality of the resulting CFD mesh. After solving the RBF problem, the displacement field for each mode is evaluated across the nodal positions of the CFD mesh, resulting in a comprehensive RBF solution database that fully maps the structural eigenvectors onto the fluid mesh. The final stage of the proposed strategy involves iterative FSI cycles aimed at deforming the baseline CFD model to balance internal structural stresses against aerodynamic loads. Each FSI cycle comprises:

1. a CFD simulation;
2. calculation of modal coordinates;
3. mesh morphing.

The FSI cycle continues until user-defined convergence criteria are met, such as when the relative difference between successive iterations in surface forces or displacements falls below a specified threshold. The CFD solver calculates the aerodynamic forces acting on the structure, such as pressure and viscous stresses on the wet surface causing deformation. To assess the displacements within the modal space, modal forces are computed using Eq. (6) as

$$N_m = \sum_{i=1}^{M} u_{m_i} f_i \tag{14}$$

Here, the modal force N_m for each structural mode is a scalar obtained by integrating the external load field across the structure, aggregating nodal fluid loads f_i projected onto the corresponding modal displacements u_{m_i}. Using Eq. (7), modal coordinates for each structural mode are then calculated:

$$\eta_m = \frac{N_m}{\omega_m^2} \tag{15}$$

The final step of the FSI cycle involves mesh morphing of the CFD mesh through a linear superposition of all selected mode shapes (RBF solutions), weighted by their respective modal coordinates as described by Eq. (14):

$$r_{CFD} = r_{CFD_0} + \sum_{m=1}^{N} \eta_m u_m \tag{16}$$

In this equation, r_{CFD_0} and r_{CFD} represent the CFD mesh node position before and after mesh morphing, respectively, and u_m is the structural eigenvector interpolated onto the CFD mesh via RBF. After the mesh modification, the aerodynamic loads on the wing surface are altered, necessitating a new CFD run at the start of each FSI cycle. The FSI process iterates until the convergence criteria are satisfied. Compared to the two-way FSI approach, this method offers significant advantages. Only one CSM computation is required for modal analysis before the FSI cycles, avoiding the computationally expensive load mapping at each cycle. Additionally, since deformation fields for each modal shape are precomputed, structural displacements do not need to be synchronized with the CFD mesh during the iterative process. These benefits result in a faster FSI computation while maintaining accuracy, provided an appropriate modal base is selected [37]. However, this approach has some drawbacks, primarily in selecting an appropriate modal truncation and the limitations of the linear theory. Referring to this latter point it is worth to mention that various methods for incorporating non-linear modes are available in the literature [39–41].

4 Aircraft Design Application

The aeroelastic analysis of the wing of the P180 Avanti EVO aircraft by Piaggio Aerospace (Fig. 4) is the industrial application employed to test the proposed approach.

The P180 [42] is a twin turboprop engine executive aircraft, with engines in pushing configuration. It is the only three-surface commercial aircraft in the world having a canard and a conventional tailplane. This configuration allows a larger cabin with a main wing located in a central position with respect to the fuselage. The Avanti EVO is an evolution of the P180, whose several improvements involve aerodynamics, avionics and propulsion systems.

Fig. 4 Piaggio P180 Avanti EVO ©PiaggioAerospace

The CFD numerical domain used for the FSI analysis consisted of a hybrid prism/tetra unstructured mesh with approximately 21 million cells. The analysis was performed in cruising condition.

To account for the engines operative conditions, appropriate boundary conditions were applied at the nacelle inlet and outlet. The results of the baseline simulation are illustrated in Fig. 5. Numbers regarding the numerical setup and results are not shown on pourpose. After obtaining the CFD solution for the baseline configuration, a hot restart approach was utilized. This involved using the converged solution from the baseline as the initial condition for all subsequent FSI calculations, thereby reducing computational time.

Figure 6 on the left shows the structural model superposed on the CAD geometry of the aircraft. Winglets are modelled in full detail including composite layups, while other parts of the model are simplified with elements of reduced complexity or fewer

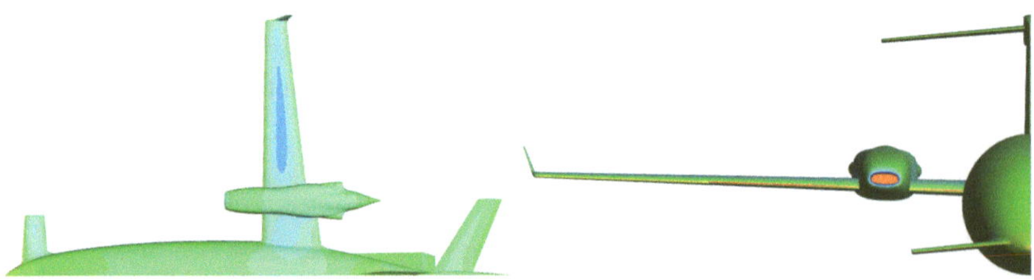

Fig. 5 C_p results obtained for the baseline configuration of the P180 model

Fig. 6 Left: Piaggio P180 structural model employed for FSI computation, right: CFD model

degrees of freedom. All the components with limited structural influence such as flaps, ailerons and spoilers are included as concentrated masses. In the aerospace industry this kind of structural model is very common, since it allows to have a reliable and accurate structural representation balancing a high level of detail on the most important districts and a reduced complexity on the less important parts. The difficulty of carrying a two-way FSI simulation is evident by comparing this simplified structural model to the CFD one, shown in Fig. 6 on the right. Notable discrepancies include the engine area, where the nacelles and the engines were represented using bar elements: in this district there are no overlapping boundaries, making it impossible to exchange loads and displacements back and forth between CSM and CFD simulations for a two-way approach.

The CSM model comprises 62,800 nodes and 79,000 elements, including plates, beams, and laminates for the wings, as well as rigid elements, rods, beams, and springs for other structural components. 30 modal shapes were initially extracted from the CSM model. Following an initial analysis, many of these modes were discarded as they pertained to local vibrations of plates not located in wetted areas thus not participating in the fluid-structure interaction. To reduce the number of modes, while faithfully reproducing the structural behaviour, a modal truncation approach was carried reducing the number of eigenvectors until a given threshold error was reached. Modes were used to define a shape parameter for each, following the baseline shape modification procedure outlined in Sect. 3.3. Figure 7 illustrates the first extracted modal shape, showing its transfer from the FEM model on the left to the CFD mesh on the right.

4.1 *FSI Model*

As it was outlined in the previous section, in the mode superposition approach the transfer of modal shapes and frequencies serves as the sole link between the CFD and CSM solvers. To accurately interpolate displacements on the wing, initially known only at structural nodes, the first step involved generating RBF source points exclusively from the CSM nodes. By setting up an initial RBF problem with only

Fig. 7 First structural modal shape and transfer to the CFD mesh

structural displacements and omitting fixed source points, a smoother mesh deformation on the wing was achieved, effectively propagating structural displacements also for the areas in which there was a gap or missing geometrical representation.

The interpolated displacements on all CFD surface nodes were subsequently used as input, along with the fixed set, for the final RBF problem. Figure 8 illustrates, for the first vibrational mode, this process, that was for each modal shape. The left image shows the RBF problem generated solely from FEM nodes, highlighting the initial (black points) and final position (green points) of the source points. The centre image displays the interpolated displacements on the CFD skin surfaces, combined with the fixed set (red points), which were used to generate the shape parameters. The resulting CFD deformation for unitary amplification is shown on the right.

The FSI model, enhanced with modal shapes and frequencies from CSM, was activated with the computational mesh being updated according to the computed modal coordinates every 100 iterations. A convergence criterion was established, defined by a relative displacement of 1 mm at the wingtip between successive iterations. Convergence was reached after 10,000 iterations, with each iteration taking 3 s. The total computation time was 30,000 s using a workstation with two Intel Xeon E5 2680 v2 with 10 cores each at 3.5 GHz and 128 Gb RAM.

Figure 9 compares the deformation resulting from the FSI analysis with the baseline geometries of the outer wing.

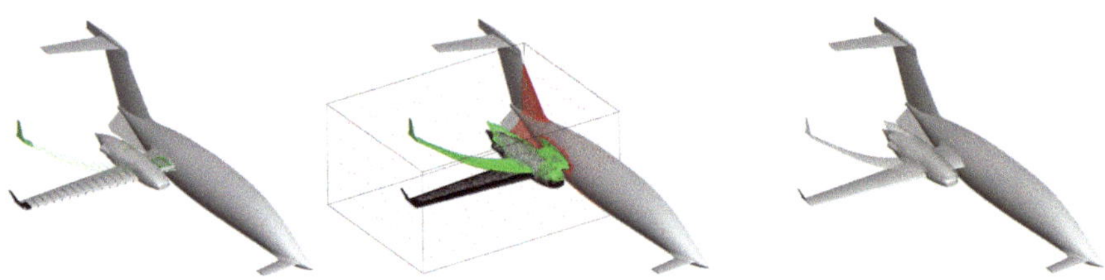

Fig. 8 RBF problem setup for the first modal shape. Left: RBF problem with FEM points, center: problem with fixed and moving sets, right: resulting CFD deformation

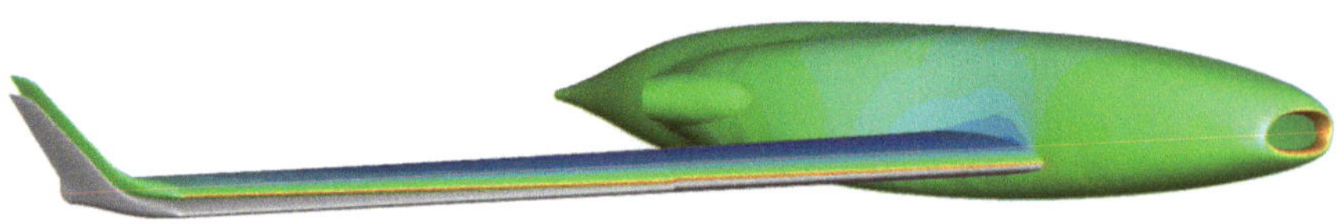

Fig. 9 Rigid–flexible model comparison for the baseline configuration

5 Conclusions

In this paper, we tackled the challenge of data transfer in FSI across models from distinct computational environments with varying levels of resolution: high accuracy for the CFD and a coarser, yet functional, numerical modelling with heterogeneous element types for the CSM. The proposed approach utilized the mode superposition method in combination with RBFs interpolation to efficiently update the CFD model, using the industrial case of the P180 Avanti EVO aircraft as a test scenario. Vibrational modes and frequencies were extracted and stored from an initial CSM analysis, creating an integrated block that combines CFD software with RBF-based mesh morphing. This setup allowed for seamless CFD computation and dynamic updates of the numerical model in response to aerodynamic loads. RBFs provided the flexibility needed to address shape and mesh discrepancies, which were significant in this application.

References

1. Bungartz, H.J., Schäfer, M.: Fluid-Structure Interaction: Modelling, Simulation, Optimisation, vol. 53. Springer Science & Business Media (2006)
2. Morand, H.J.P., Ohayon, R.: Fluid Structure Interaction-Applied Numerical Methods. Wiley (1995)
3. Dowell, E.H., Hall, K.C.: Modeling of fluid-structure interaction. Annu. Rev. Fluid Mech. **33**(1), 445–490 (2001)
4. Hou, G., Wang, J., Layton, A.: Numerical methods for fluid-structure interaction—a review. Commun. Comput. Phys. **12**(2), 337–377 (2012)
5. Benra, F.K., Dohmen, H.J., Pei, J., Schuster, S., Wan, B.: A comparison of one-way and two-way coupling methods for numerical analysis of fluid-structure interactions. J. Appl. Math. **2011** (2011)
6. Hübner, B., Walhorn, E., Dinkler, D.: A monolithic approach to fluid-structure interaction using space-time finite elements. Comput. Methods Appl. Mech. Eng. **193**(23–26), 2087–2104 (2004)
7. Degroote, J.: Partitioned simulation of fluid-structure interaction. Arch. Comput. Methods Eng. **20**(3), 185–238 (2013)
8. Wang, T., Wüchner, R., Sicklinger, S., Bletzinger, K.U.: Assessment and improvement of mapping algorithms for non-matching meshes and geometries in computational FSI. Comput. Mech. **57**(5), 793–816 (2016)
9. Biancolini, M., Chiappa, A., Giorgetti, F., Groth, C., Cella, U., Salvini, P.: A balanced load mapping method based on radial basis functions and fuzzy sets. Int. J. Numer. Meth. Eng. **115**(12), 1411–1429 (2018)
10. Cebral, J.R., Lohner, R.: Conservative load projection and tracking for fluid-structure problems. AIAA J. **35**(4), 687–692 (1997)
11. Bazilevs, Y., Takizawa, K., Tezduyar, T.E.: Computational Fluid-Structure Interaction: Methods and Applications. Wiley (2013)
12. Keye, S.: Fluid-structure coupled analysis of a transport aircraft and flight-test validation. J. Aircr. **48**(2), 381–390 (2011)
13. Groth, C., Cella, U., Costa, E., Biancolini, M.: Fast high fidelity CFD/CSM fluid structure interaction using RBF mesh morphing and modal superposition method. Aircr. Eng. Aerosp. Technol. **91**(6), 893–904 (2019). https://doi.org/10.1108/AEAT-09-2018-0246

14. Jakobsson, S., Amoignon, O.: Mesh deformation using radial basis functions for gradient-based aerodynamic shape optimization. Comput. Fluids **36**(6), 1119–1136 (2007). https://doi.org/10.1016/j.compfluid.2006.11.002
15. de Boer, A., van der Schoot, M.S., Bijl, H.: Mesh deformation based on radial basis function interpolation. Comput. Struct. **85**(11–14), 784–795 (2007). https://doi.org/10.1016/j.compstruc.2007.01.013
16. van Zuijlen, A.H., de Boer, A., Bijl, H.: Higher-order time integration through smooth mesh deformation for 3D fluid–structure interaction simulations. J. Comput. Phys. **224**(1), 414–430 (2007). https://doi.org/10.1016/j.jcp.2007.03.024
17. Biancolini, M., Cella, U.: An advanced RBF morph application: coupled CFD CSM aeroelastic analysis of a full aircraft model and comparison to experimental data. In: GB, pp. 243–258 (2010)
18. Groth, C., Costa, E., Biancolini, M.E.: RBF-based mesh morphing approach to perform icing simulations in the aviation sector. Aircr. Eng. Aerosp. Technol. **91**(4), 620–633 (2019)
19. Papoutsis-Kiachagias, E., Andrejasic, M., Porziani, S., et al.: Combining an RBF-based morpher with continuous adjoint for low-speed aeronautical optimization applications. In: ECCOMAS Congress 2016—Proceedings of the 7th European Congress on Computational Methods in Applied Sciences and Engineering (2016)
20. Kapsoulis, D., Asouti, V., Giannakoglou, K., et al.: Evolutionary aerodynamic shape optimization through the RBF4AERO platform. In: Proceedings of the VII European Congress on Computational Methods in Applied Sciences and Engineering (ECCOMAS Congress 2016). Institute of Structural Analysis and Antiseismic Research School of Civil Engineering National Technical University of Athens (NTUA), Athens, Greece, pp. 4146–4155 (2016)
21. Biancolini, M., Viola, I., Riotte, M.: Sails trim optimisation using CFD and RBF mesh morphing. Comput. Fluids **93**, 46–60 (2014)
22. Van Zuijlen, A., de Boer, A., Bijl, H.: Higher-order time integration through smooth mesh deformation for 3D fluid–structure interaction simulations. J. Comput. Phys. **224**(1), 414–430 (2007)
23. Domenico, N., Groth, C., Wade, A., Berg, T., Biancolini, M.: Fluid structure interaction analysis: vortex shedding induced vibrations. Procedia Struct. Integr. **8**, 422–432 (2018). https://doi.org/10.1016/j.prostr.2017.12.042
24. Biancolini, M.E., Cella, U., Groth, C., Chiappa, A., Giorgetti, F., Nicolosi, F.: Progresses in fluid-structure interaction and structural optimization numerical tools within the EU CS RIBES project, pp. 529–544. Springer (2019)
25. Biancolini, M., Cella, U., Groth, C., Genta, M.: Static aeroelastic analysis of an aircraft wind-tunnel model by means of modal RBF mesh updating. J. Aerosp. Eng. **29**(6) (2016). https://doi.org/10.1061/(ASCE)AS.1943-5525.0000627
26. Andrejašič, M., Eržen, D., Costa, E., Porziani, S., Biancolini, M., Groth, C.: A mesh morphing based FSI method used in aeronautical optimization applications. In: ECCOMAS Congress 2016—Proceedings of the 7th European Congress on Computational Methods in Applied Sciences and Engineering (2016)
27. de Boer, A., van Zuijlen, A.H., Bijl, H.: Review of coupling methods for non-matching meshes. Comput. Methods Appl. Mech. Eng. **196**(8), 1515–1525 (2007)
28. Thomson, W.: Theory of Vibration with Applications. CRC Press (2018)
29. Cook, R.D.: Concepts and Applications of Finite Element Analysis. Wiley (2007)
30. Meirovitch, L.: Fundamentals of Vibrations. McGraw-Hill (2001)
31. Micchelli, C.A.: Interpolation of scattered data: distance matrices and conditionally positive definite functions. Constr. Approx. **2**(1), 11–22 (1986). https://doi.org/10.1007/BF01893414
32. Beckert, A., Wendland, H.: Multivariate interpolation for fluid-structure-interaction problems using radial basis functions. Aerosp. Sci. Technol. **5**(2), 125–134 (2001). https://doi.org/10.1016/S1270-9638(00)01087-7
33. Cella, U., Biancolini, M.E.: Aeroelastic analysis of aircraft wind-tunnel model coupling structural and fluid dynamic codes. J. Aircr. **49**(2), 407–414 (2012)

34. Andrejašič, M., Eržen, D., Costa, E., Porziani, S., Biancolini, M., Groth, C.: A mesh morphing based FSI method used in aeronautical optimization applications (2016)
35. Heeg, J.: Overview of the aeroelastic prediction workshop, p. 783 (2013)
36. Groth, C., Porziani, S., Chiappa, A., et al.: Structural validation of a realistic wing structure: the RIBES test article. Procedia Struct. Integr. **12**, 448–456 (2018)
37. Cella, U., Biancolini, M.E.: Aeroelastic analysis of aircraft wind-tunnel model coupling structural and fluid dynamic codes. J. Aircr. **49**(2), 407–414 (2012). https://doi.org/10.2514/1.C031293
38. Groth, C.: Adjoint-based shape optimization workflows using RBF. Ph.D. thesis. University of Rome Tor Vergata (2015). https://doi.org/10.13140/RG.2.2.33913.06245
39. Liu, J.: Exact solution of nonlinear hysteretic responses using complex mode superposition method and its application to base-isolated structures. J. Eng. Mech. **131**(3), 282–289 (2005)
40. Au, F., Yan, Z.: Dynamic analysis of frames with material and geometric nonlinearities based on the semirigid technique. Int. J. Struct. Stab. Dyn. **8**(03), 415–438 (2008)
41. Lamarque, C.H., Janin, O.: Modal analysis of mechanical systems with impact non-linearities: limitations to a modal superposition. J. Sound Vib. **235**(4), 567–609 (2000)
42. P180 Avanti EVO: http://www.avantievo.piaggioaerospace.it. Accessed: 2019-09-17

Digital Twin Concept Application: The MOthIF Case

Giorgio Travostino

Abstract Under the CLEANSKY 2 framework SONACA and Von Karman Institute of Fluid Dynamics, establish the Research and Innovation Action MOTHIF (MOdel Testing of HIgh lIFt model) consortium with Piaggio Aerospace as Topic Leader. Funded by the European Commission under the Horizon 2020 program, it was in charge to validate with wind tunnel testing a selected high lift device system for Small Air Transport aircraft (SAT). The wind tunnel tests have been performed in the open-jet test section of the Von Karman Institute of Fluid Dynamics low-speed wind tunnel L1-A. The wing model, representing a blown flap airfoil in which an air jet from the main trailing edge skims the flap upper side, was very large due to the accommodation of the pneumatic feeding into the trailing edge main airfoil. The ratio between the test chamber diameter and the test model large dimension necessarily led to blockage effect. The Digital Twin approach based on CFD (Computational Fluid Dynamic) computation helped the identification of the aerodynamic interference between wind tunnel and the airfoil model and significantly contributed to the test success.

1 Introduction

CleanSky 2 SAT (Small Air Transport) research program has been focused on more electric technology able to reduce fuel burnt and the emissions, also reducing operative costs with the target to meet the emerged demand for low fare thin haul regional routes. Performing regional missions and reducing emissions led to consider short airfields not exceeding 800 m and aircraft configurations with reduced surfaces and high lift capabilities wings. Reducing cost led to find simple very effective STOL (Short Take Off and Landing) solution.

Within the framework of SAT a preliminary study, aiming at improve the take-off and landing aircraft performance, was carried out by CIRA (Centro Italiano Ricerche

G. Travostino (✉)
Baykar Piaggio Aerospace S.p.A., Villanova D'Albenga, Savona, Italy
e-mail: gtravostino@piaggioaerospace.it
URL: http://www.piaggioaerospace.it

M. E. Biancolini et al. (eds.), *Digital Twins for Aerospace*, Springer Aerospace Technology, https://doi.org/10.1007/978-3-032-05022-9_8

Fig. 1 SAT Green 19-seater E-STOL EIS2032 aircraft

Aerospaziali, Italy) and ILOT (Lukasiewicz Institute of Aviation, Poland) on two dimensional wing sections, the first to define an optimization of standard solutions for the P180 aircraft, the second to verify the feasibility of blown flap approach for the PZL-Mielec M28 aircraft.

At the end of the study, a down selection indicated the blown flap concept as preferable for a new airfoil designed for SAT Green 19-seater E-STOL EIS2032 aircraft (Fig. 1).

A following wind tunnel test provided the verification of the proof of concept design performances.

2 Airfoil GP7

The design requirements for the airfoil was to reach the maximum Lift Coefficient in clean configuration (flap retracted) in the following flow condition:

Mach number = 0.305 (100 m/s)
Reynolds number = 5,345,908 (for an airfoil chord = 1 m)

With the following geometrical constraint:

Percentage Maximum Thickness $\geq$ 13%.

With the aerodynamic target to maintain the same Efficiency of the starting design airfoil at

Lift Coefficient: 0.6 (High Speed Cruise)
Lift Coefficient: 0.95 (Long Range Cruise)

and the Cruise Pitching Moment > − 0.05.

The basic concept that inspired the airfoil choice was to hold high lift capabilities in case of blowing system failure. This constraint led to consider a fowler flap configuration that shifted from the traditional concept of COANDA aerodynamic effect [1] to the airfoil boundary layer control.

The starting airfoil for the design process was the NACA 43013 that guaranteed the maximum Lift Coefficient equal to 1.9.

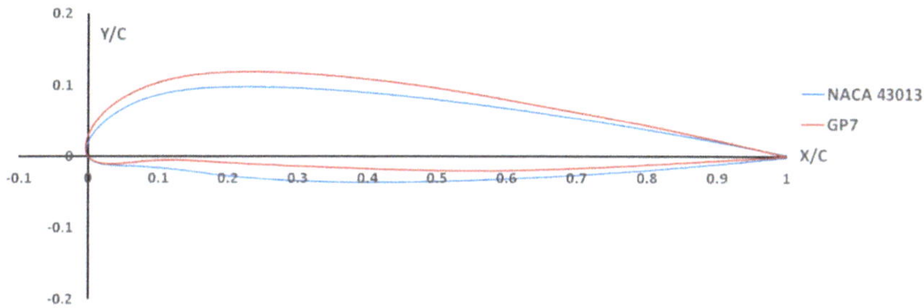

Fig. 2 Airfoils shapes

Table 1 Geometry comparison between the starting and optimized airfoils

% chord	NACA 43013	GP7
Maximum thickness	13.01	13.42
Maximum thickness @x/c%	29.94	27
Maximum camber	3.68	4.56
Maximum camber @x/c%	14.60	16.85

The chosen airfoil passed an optimization cycle using a Bi-Dimensional Multi Objective Gradient-Based Optimizer coupled with approximated Bezier parametrization technique and based on the XFOIL [2] aerodynamic solver, with the objective to increase high lift, holding the long-range cruise performances of the original airfoil.

The optimization led to a thicker and more cambered airfoil moving towards the trailing edge both the maximum camber value and toward the leading edge the maximum thickness value. Figure 2 presents the airfoils geometries scaled by the airfoil chord and Table 1 shows the geometric quantitative comparison between the two airfoils.

The analysis via Reynolds Averaged Navier Stokes equations CFD++ solver [3, 9] k-ω SST Langtry-Menter free boundary layer transition turbulence model [4, 5] certified the design process, indicating a maximum Lift Coefficient (CL) gain of 12% holding the Long Cruise Efficiency. The high lift gain payed an increase of Pitching Moment Coefficient (Cm), considered acceptable as presented in Figs. 3 and 4.

The housing of the blow jet duct in the trailing edge, led to the definition of a new rear part of the airfoil. XFOIL panel code [2] was the elected tool to comply with the target.

Figure 5 shows the aerodynamic effect of this geometry modification consisting in a slight worsening of the airfoil performances. In particular, there is a slight reduction in the maximum lift coefficient and the efficiency of long-range cruise.

The graphs related to moment coefficient Cm and efficiency E presented in Fig. 5 show non-linear traits starting from an angle of incidence of 3.5°.

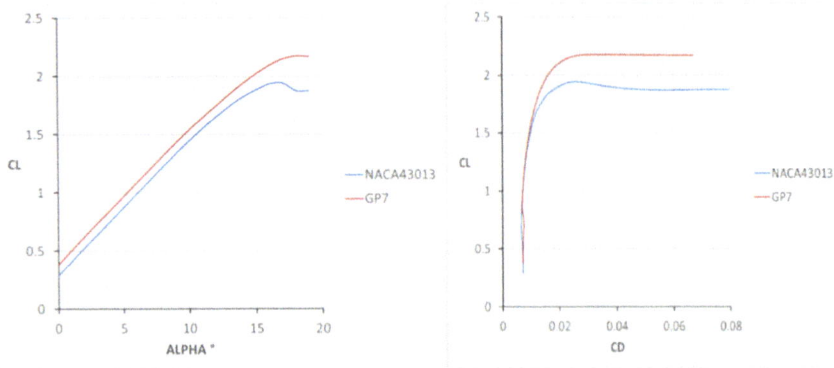

Fig. 3 Airfoils aerodynamic coefficients curve comparison. Left side: lift coefficient versus ALPHA (Angle of attack) curves. Right side: polar curves (Lift vs. drag coefficient)

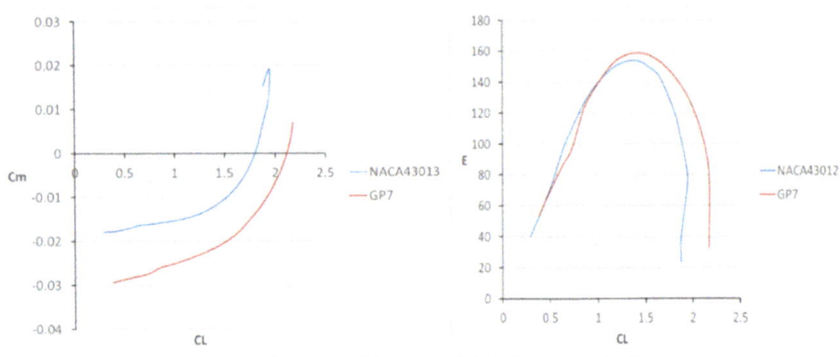

Fig. 4 Airfoils aerodynamic coefficients curve comparison. Left side: pitching moment versus lift coefficient curves. Right side: efficiency versus lift coefficient

The analysis of the XFOIL code results indicates that from this incidence, the transition of the boundary layer from laminar to turbulent state shifts from 10 to 77% on the profile's underside.

This is therefore a prediction of boundary layer re-laminarization. This phenomenon occurs when the boundary layer experiences a favorable pressure gradient that accelerates the flow and, by reducing the size of turbulent structures, allows it to reorganize into a laminar state (Fig. 6).

The application of blowing duct was borrowed from ILOT preliminary studies and not optimized for the present airfoil. Figure 7 shows the two-dimensional geometry with the added fowler flap.

The planned two-dimensional analysis considered four flap settings and provided optimized flap gaps and overlaps with the aim to hold acceptable airfoil performance in blow off condition. Table 2 summarizes the RANS analysis results on the defined geometry in flapped configuration varying the flap setting and the blow Pressure

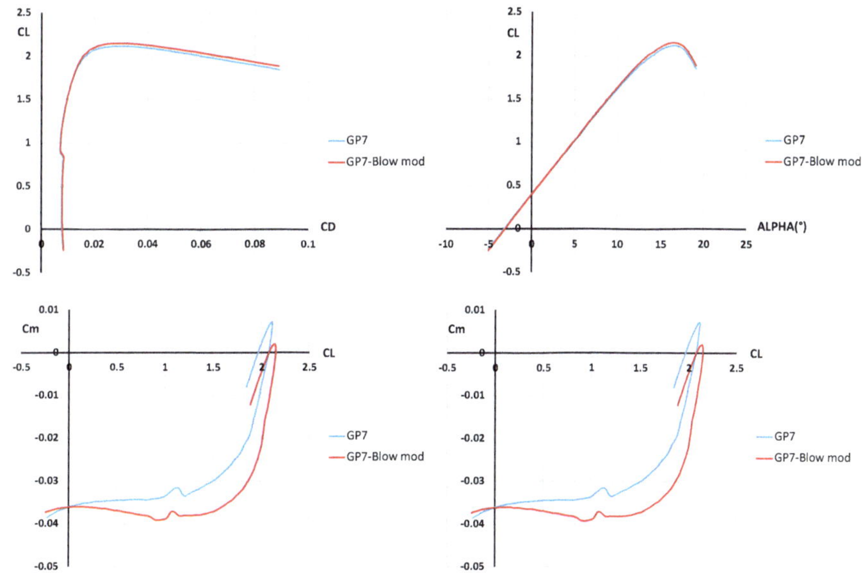

Fig. 5 Airfoils aerodynamic coefficients curve comparison. Top left: polar curves. Top right: lift coefficient versus ALFA (Angle of attack). Bottom right: pitching moment versus lift coefficient and efficiency (CL/CD) versus angle of attack curves

Fig. 6 Boundary layer laminar-turbulent transition: position versus lift coefficient

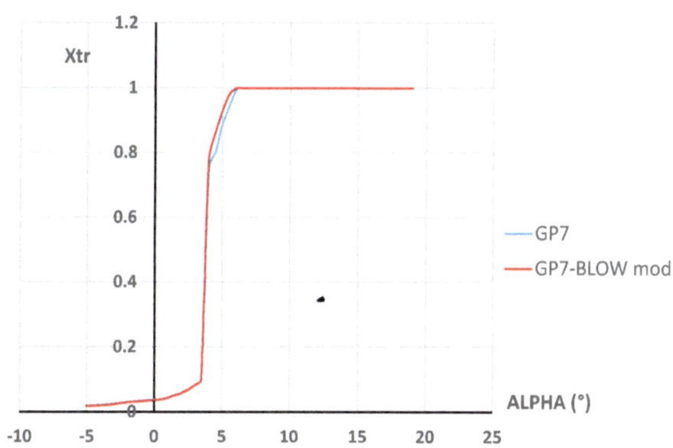

Ratio (p/p_0, the ratio between the pressure of the supplied air to the blow duct and the ambient pressure).

Figure 8 shows the graphs from the tabulated data. The FLAP 30 configuration was the most performing set, allowing for reaching the maximum Lift Coefficient of 4.77 at the Pressure Ratio $p/p_0 = 1.88$, which generated the blow duct sonic throat. FLAP 40 showed a performance fall at Pressure Ratio equal to 1.68 due to flow separation (Fig. 9).

The two-dimensional analysis revealed that the trailing edge air blow acted more on the main airfoil than on the flap, reducing the wake thickness and increasing

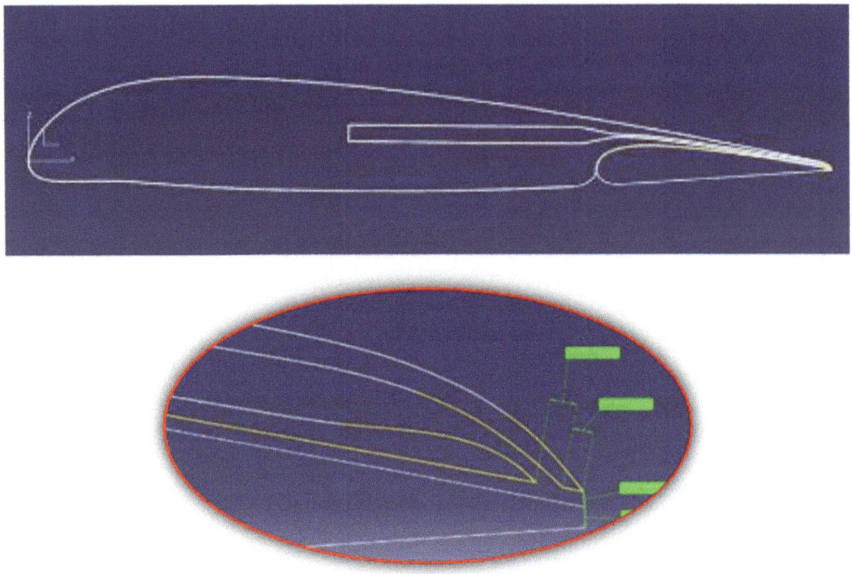

Fig. 7 Blowing trailing edge shape detail

Table 2 Flapped configuration RANS analysis. Effect of FLAP set and p/p_0

(p/p_0)/FlapRotation (°)	1	1.28	1.48	1.68	1.88
15	3.21	3.62	3.84	/	/
20	3.41	3.83	4.08	4.27	4.42
30	3.69	4.23	4.44	4.62	4.77
40	3.86	4.37	4.53	4.42	4.54

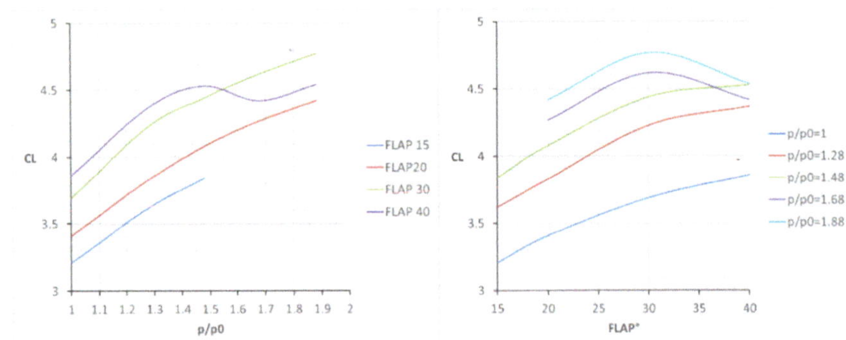

Fig. 8 Effect on CL of pressure ratio (p/p_0) on flap setting and of flap setting on pressure ratio

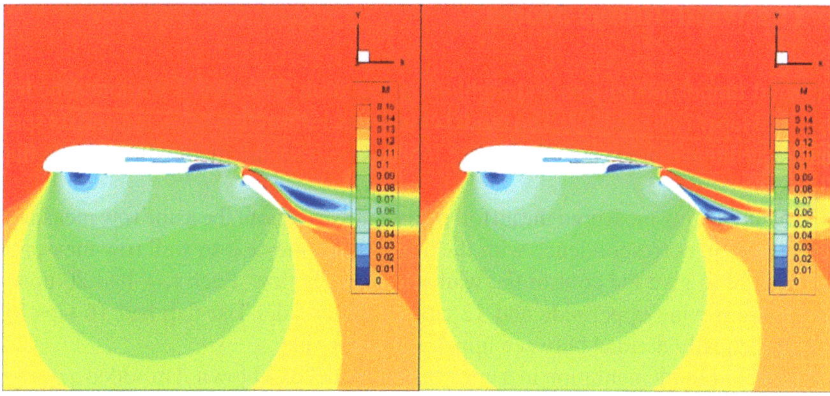

Fig. 9 FLAP 40 maximum lift coefficient drop due to flap flow detachment: left $p/p_0 = 1.48$, right $p/p_0 = 1.68$. Mach number maps

the leading edge suction (Fig. 10), working as a boundary layer control rather than COANDA effect. From the figure, it is clearly visible that the application of blowing results in a thinning of the boundary layer, leading to an increase in circulation which produces a more extensive increase in pressure in the front part of the profile. Therefore, it is the main profile that benefits from the blowing rather than the flap.

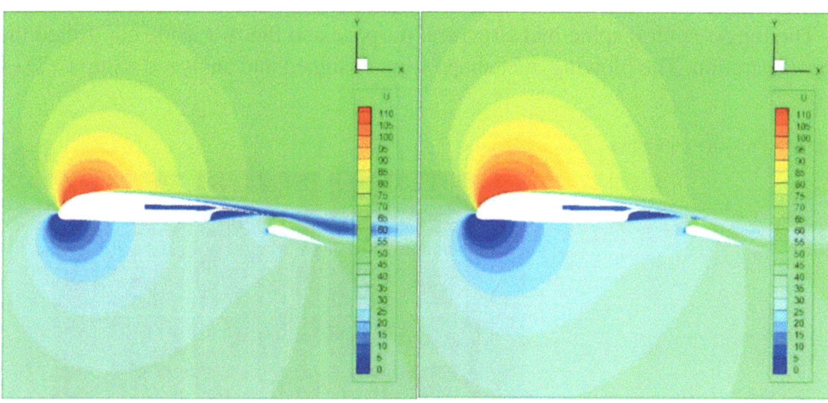

Fig. 10 Alpha = 14°. Pressure ratio = 1.48. Left—blow off. Right—blow on

3 The Wind Tunnel Model

The wind tunnel model was a straight wing based on GP7 airfoil without dihedral angle and taper ratio equal 1. The model chord was 1.35-m flap retracted and its span was 1.5 m-long. Three row of pressure tap was both on the main wing and on flap (Fig. 11).

The middle row measured the airfoil pressure distribution; the aside row measured the pressure in correspondence with the blowing jet; the next to wall row measured the pressure to give the control the wall effect on 2D flow. Each row had 20 upper taps on main wing upper side and 12 taps on lower side. The flap had 16 taps for each row equally divided between upper and lower side.

The airfoil implementation in a wind tunnel model of a blown flap added a design complication with respect to a traditional airfoil with a sharp trailing edge. In this case, the duct design assumed an important role in the model definition. Structural robustness reasons mainly drove the planform shape choice, taking in account the local extremely low thickness. The entire blowing duct of the model has been redesigned to minimize the extent of flow separations within it, while maintaining the sectional area distribution obtained by ILOT as a result of a two-dimensional study. The prevention of extensive separated flow regions along the duct, especially near the exit, aims to minimize flow instabilities and the resulting unsteadiness of the exiting jet as much as possible. The duct had central part with a rectangular longitudinal section and planform upstream tapered for the link to the air-feeding duct. The rectangular sections increased their area to reach the blown exit (Fig. 12).

The longer central spine had structural purpose and the two aside controlled the flow separation. The blowing duct shape were designed and analyzed using CFD++

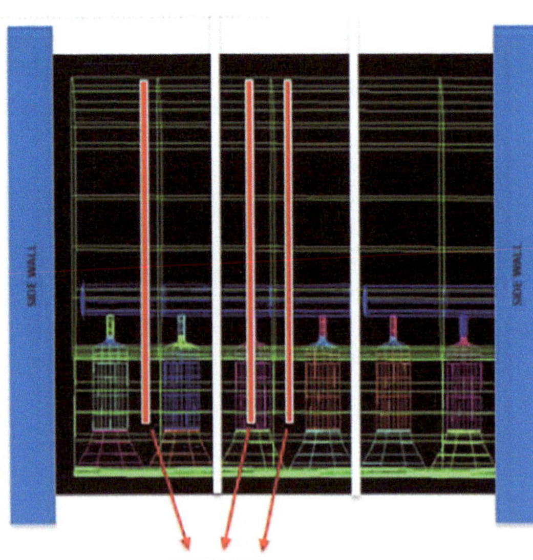

Fig. 11 Pressure taps rows

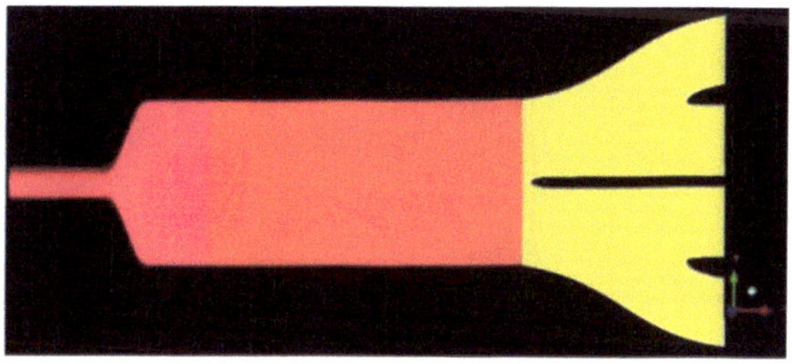

Fig. 12 Blow duct

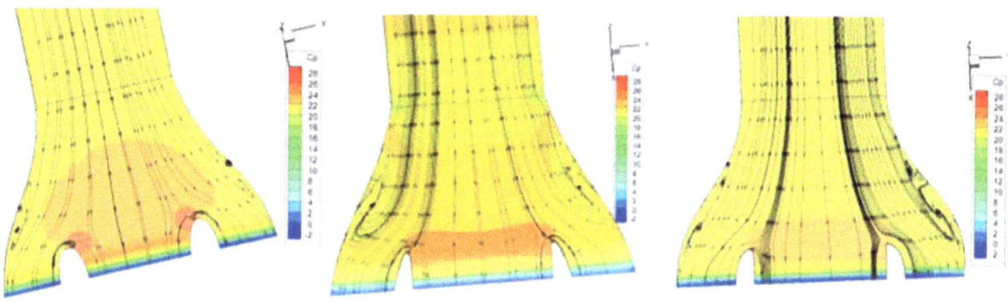

Fig. 13 Blow duct shape design

k-ω SST [3, 6] turbulence model, directly on the wind tunnel model with infinite sidewalls. The model had six trailing edge blowing modules. Figure 13 shows, in a chronological sequence from left to right, the development of the duct design, culminating in a 10% increase in the exit flow rate compared to the initial geometry on the left.

After the completed aerodynamic design began the loads computation and the structural verification. The aerodynamic loads was obtained from two CFD++, k-ω SST turbulence, computations in free air, varying Angle of Attack from 10° to 15° at Pressure Ratio 1 and 1.88, the maximum reachable in wind tunnel, respectively. The most severe loads for each model part resulted from the computations at these two different Pressure Ratio:

Main wing

- Maximum normal force from maximum lift at maximum Pressure Ratio
- Maximum pitching moment at maximum Pressure Ratio
- Maximum longitudinal force from maximum lift at maximum Pressure Ratio.

Flap

- Maximum normal force from maximum flap deflection and blow off
- Maximum pitching moment from maximum flap deflection blow off
- Maximum longitudinal force from maximum lift at maximum Pressure Ratio.

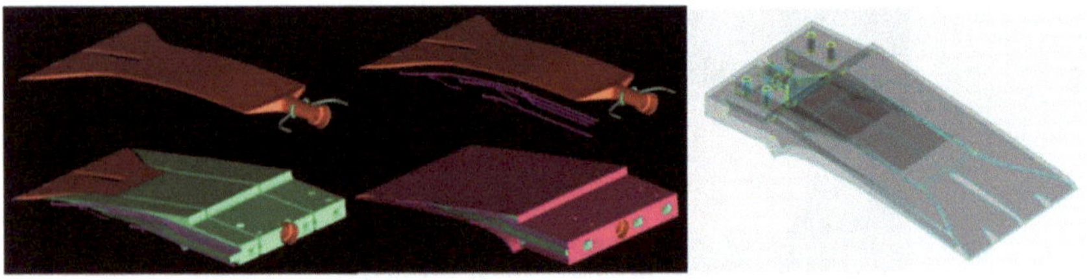

Fig. 14 Blow duct design: final geometry

On this basis, SONACA and Piaggio Aerospace started the model structural design. SONACA was in charge of model manufacturing using machining technique. Piaggio Aerospace was in charge of main wing trailing edge manufacturing. This part (Fig. 14), carrying the thin air blow duct and the piping for pressure taps, was built by Piaggio Aerospace with the collaboration of 3D-Systems S.R.L using the stereo-lithography technique with Accura Bluestone material.

In this late case, the coupling CFD with structural analysis allowed for the reaching of two main goal:

- wind tunnel model safety load factor reduction (accepted by Von Karman Institute of Fluid Dynamics) from 5 to 3.5
- The reduction of Accura Blue Stone quantity employed nearly 49% with local deformation inside the tolerances.

Figures 15 and 16 show the model and the air supply air piping located in the main wing.

Fig. 15 Wind tunnel model

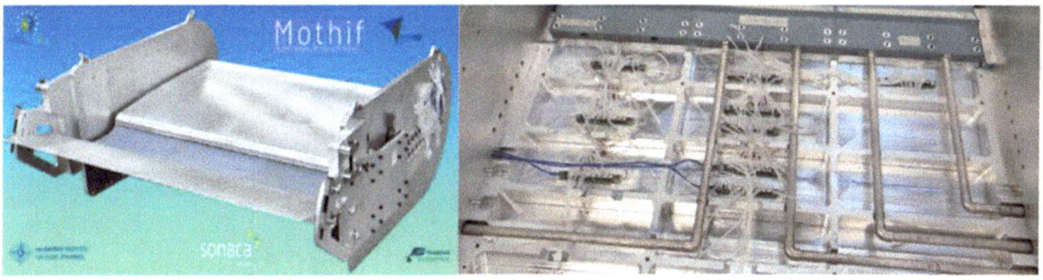

Fig. 16 Wind tunnel model

4 The Wind Tunnel

The selected wind tunnel for the tests execution was the Von Karman Institute of Fluid Dynamics low-speed open-jet test section wind tunnel L1-A (Fig. 17). The open-jet test section measures 3 m diameter and 4.5 m length. A variable speed DC motor of 580 kW drove the contra-rotating fans of this facility, allowing a continuous variation of velocity from 2 to 60 m/s. The contraction ratio is four with a typical turbulence level of 0.3%.

4.1 Model-Test Chamber Arrangement

SONACA produced the MOTHIF airfoil model. Von Karman Institute of Fluid Dynamics in collaboration with SONACA designed a dedicated support (Fig. 18).

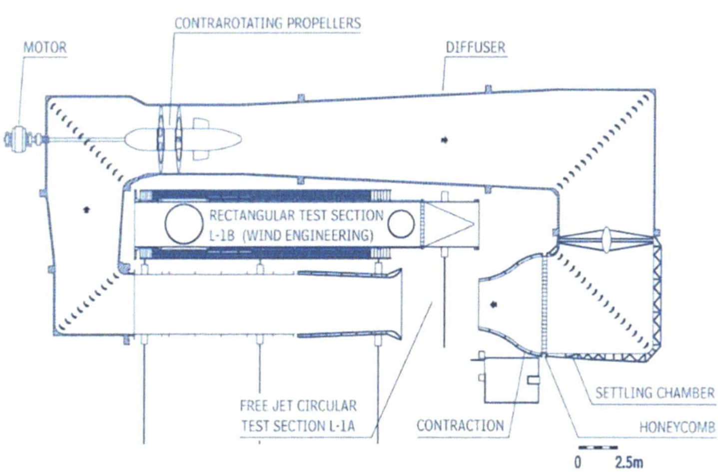

Fig. 17 L1-A wind tunnel plant

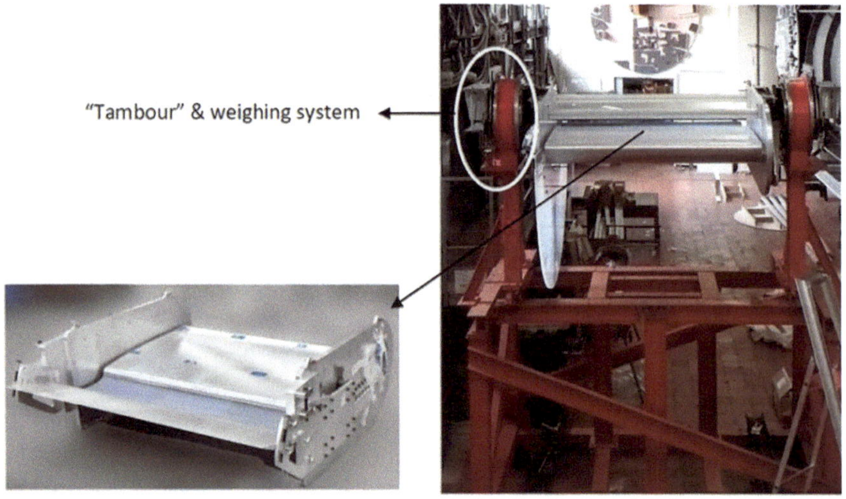

Fig. 18 Model support

The model was installed "up-side-down" to direct the lift force, generated by the model, towards the ground for safety reason.

The support allowed for holding in place the model and for the regulating its Angle of Attack. It housed the force transducers of the weighing system. A specially manufactured intermediate piece, connected to the support rotating part and named Tambour, realized the change of the model Angle of Attack. The largest as possible lateral endplates walls, together with Tambour closing fairing, reduced the aerodynamic interaction between the flow and the support, avoiding 3D flow development on the model and reducing the blockage effects (Fig. 19). Two parts composed the lateral endplates: a circular part attached to the model and thus rotating when changing the Angle of Attack and a rectangular part attached to the wind tunnel floor (Figs. 20 and 21).

The circular section of the end plates had a diameter of 2.4 m and was made of wood. A part of the endplate was made of Plexiglas to allow for flow visualizations (Fig. 20).

The rectangular part of the lateral end plates was 4.25 m long and 3.4 m high, installed flush with the circular section (Fig. 21), not contributing to the measured drag.

A manifold supplied the compressed air to the six trailing edge blowing modules of the MOTHIF wing model (Fig. 22). The manifold was composed by the main pressure inlet, a mass flowmeter, a combination of six ball valves and pressure regulators. The pressure regulators controlled independently the inlet pressure of each trailing edge blowing module. A flexible pipe connected the manifold outlets to their respective trailing edge blowing modules.

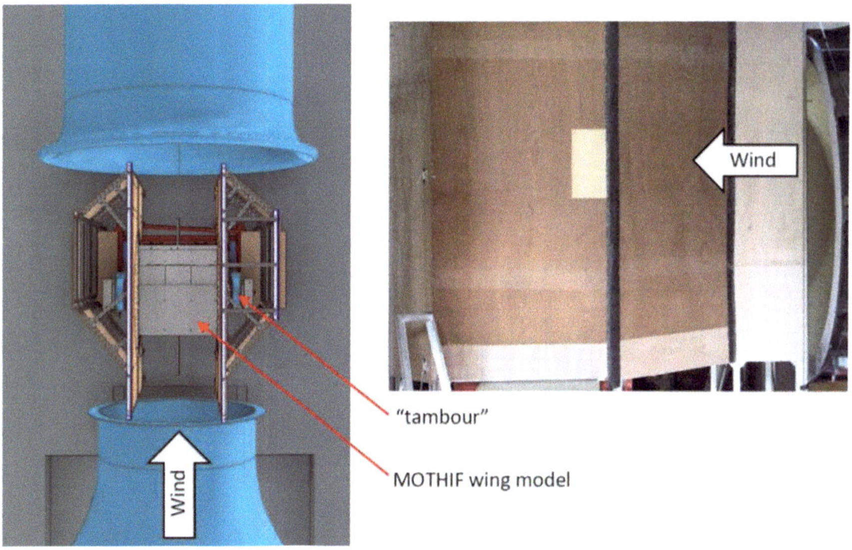

Fig. 19 L1-A wind tunnel test chamber external side wooden fairings

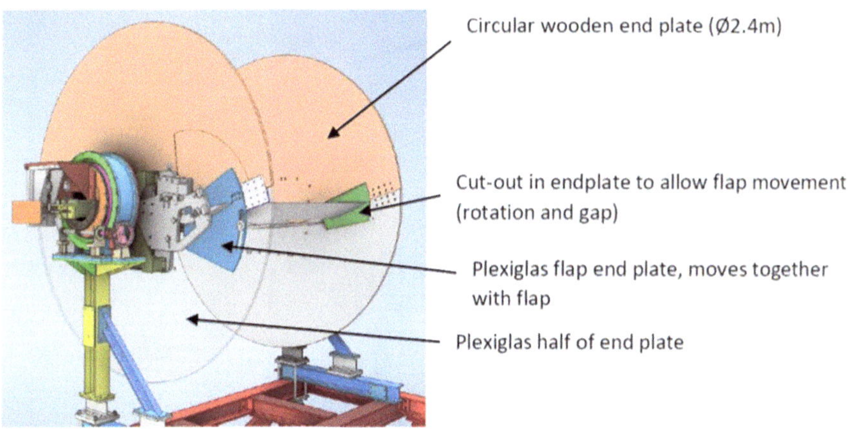

Fig. 20 L1-A wind tunnel test chamber circular endplates

4.2 Trailing Edge Blowing System

A vortex flowmeter, with a 1% accuracy and installed on the manifold, controlled the air mass flow rate supplied to the blowing duct. The sensor used three primary sensing elements for flow measurement: a vortex-shedding velocity sensor, a temperature sensor, and a solid-state pressure transducer. The described system measured the total mass flow; dividing the total mass flow by six, the individual mass flow by each

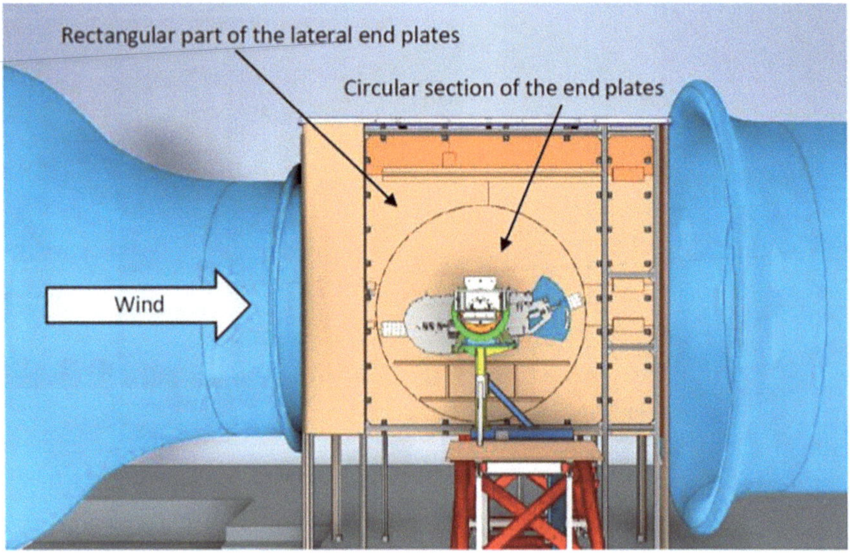

Fig. 21 L1-A wind tunnel test chamber side wooden endplates

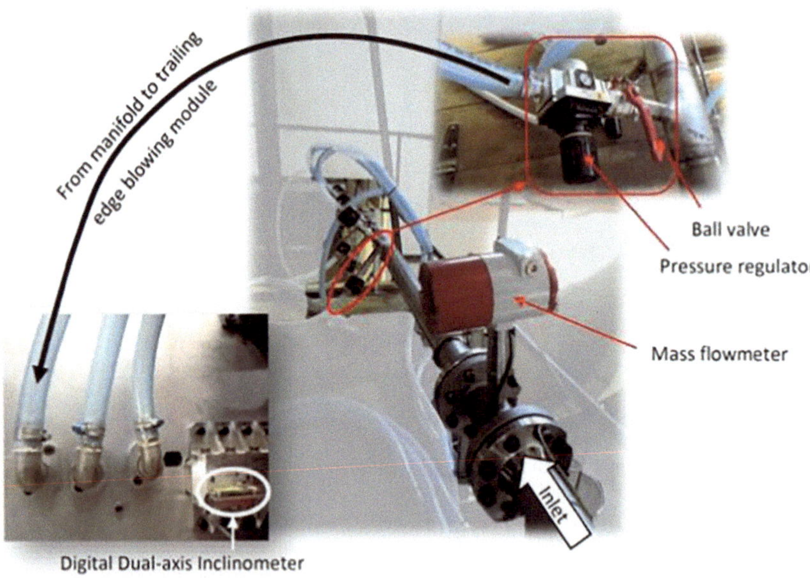

Fig. 22 Compressed air manifold for trailing edge air feeding

trailing edge module as long as the pressure of each trailing edge blowing module can be obtained.

4.3 Angle of Attack Regulation

An inclinometer, installed as in Fig. 22, controlled the Angle of Attack setting of the model.

4.4 Free Stream Velocity Measurement

Two Prandtl-tubes, installed in front of the test model, allowed for calculate the free stream wind tunnel velocity (Fig. 23). They were pneumatically connected one to each other. This is to avoid any effects on the freestream velocity due to a different Angle of Attack of the model. This instrument measured the dynamic pressure through a pressure sensor, which allows for the wind tunnel free stream velocity computation.

Fig. 23 Prandtl tubes. Wind tunnel velocity calculation

5 The Wind Tunnel Wall Aerodynamic Interference

Following the wind tunnel test chamber definition, the ratio between the test chamber section and the model thickness and span suggested a preliminary quantitative analysis on the wall aerodynamic interference acting on the model. The study considered four different sidewall dimensions, which named Small (S), Medium (M), Large (L) and Very Large (VL) (Fig. 24). The flap was 15° deflected.

All the calculations were in blow off conditions.

Figure 26 shows the calculation results, performed adopting the K-omega SST turbulence model, on a span wise cut at 330 mm from the model leading edge (Fig. 25).

The sidewall S poorly counteracted the flow roll up as showed by Fig. 26. On the left, V, y-component velocity, reveals that there were an important transversal velocity component and, on the right, Pressure Coefficient maps shows that the model tip has poor two-dimensional characteristic. This caused a change of the flow Angle of Attack, strongly reducing it by 15° (Fig. 27).

In the M configuration, the tip vortex moved away from model surface (Fig. 28 left side) and the pressure coefficient assumed a two-dimensional behavior (Fig. 28 right side).

Figure 29 shows the results on the L configuration. The tip vortices weakened and moved far away from the model surface (Fig. 29 left side). The flow cut displays a two-dimensional pressure distribution (Fig. 29 right side).

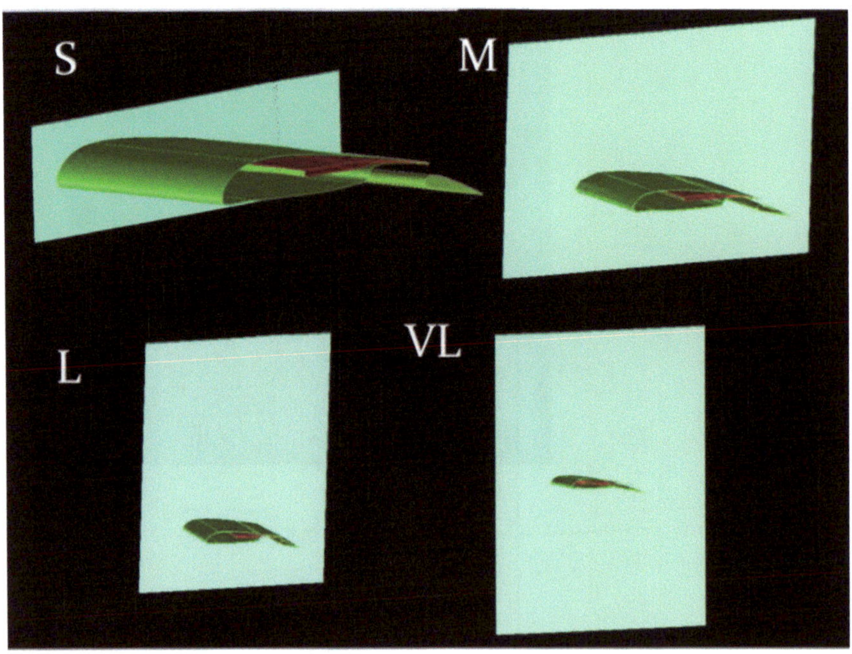

Fig. 24 Analyzed wind tunnel model sidewalls

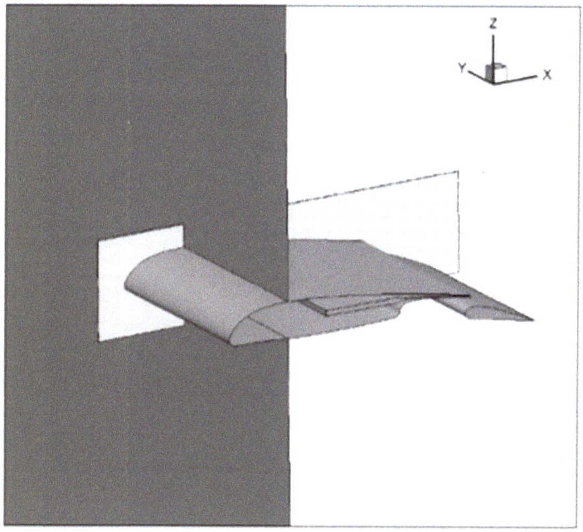

Fig. 25 Flow span wise cut at 330 mm from the model leading edge

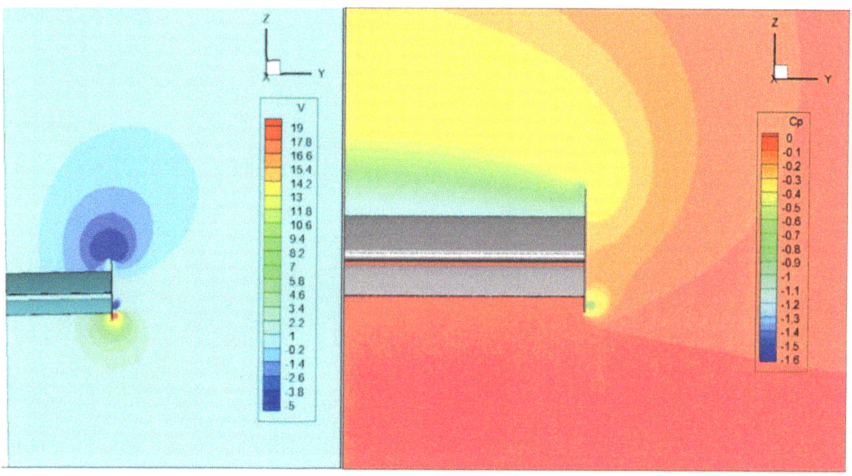

Fig. 26 The S sidewall effect

Increasing the sidewall (VL configuration), extending the surface in the lower model side also, led to the result presented in Fig. 30, which shows on the right the pressure on the sidewall facing the model. The undisturbed flow pressure in this calculation was 101,325 Pa, being the conventional altitude 0 m. On the sidewall, the difference between the outer and inner side pressure allowed for flow roll up that still had effect on the Angle of Attack of the wind tunnel model.

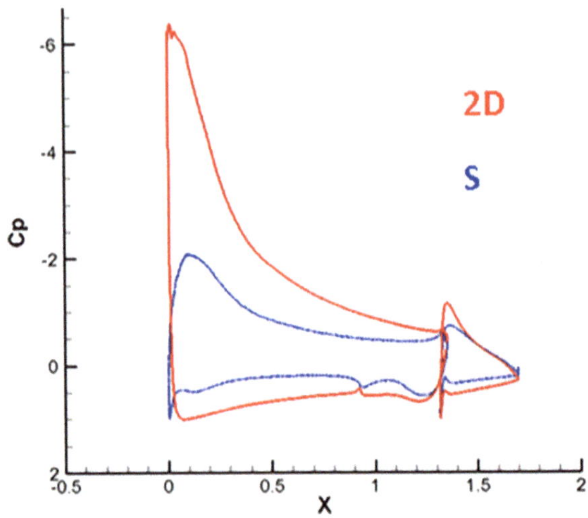

Fig. 27 The S sidewall effect—angle of attack

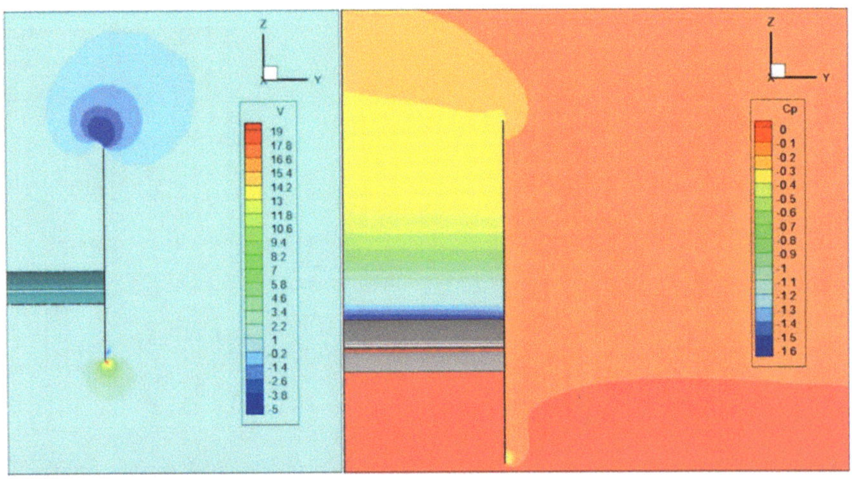

Fig. 28 The M sidewall effect

For this case and a fortiori for the previous cases it is clear that the span wise two-dimensional pressure distribution does not assure to reproduce a free air condition.

Figure 31 collects the Pressure Coefficient distributions in the centerline of the model. The sidewall enlargement effect in these graphs is clearly appreciable, considering the movement of the stagnation point away from the model leading edge to increase the flow Angle of Attack. Figure 32 presents the Lift Coefficient whose value increases enlarging the sidewalls but with a decreasing gain from Small to Very Large one.

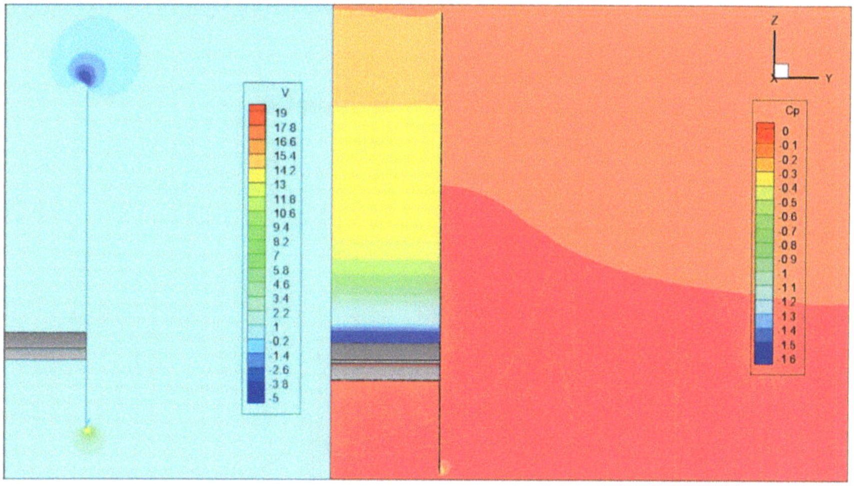

Fig. 29 The L sidewall effect

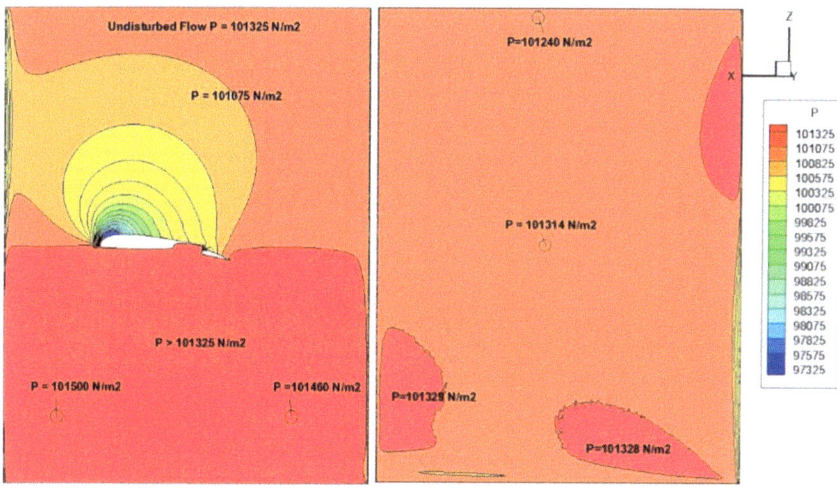

Fig. 30 The VL sidewall effect. Right: wall wind tunnel model facing. Left: outside

The analysis led to the conclusion that any sidewall set in the wind tunnel would result in some model performance penalization, unless to provide a sidewall closure.

Considering the wind tunnel assigned for the test, it seemed a proper approach to generate a wind tunnel digital twin to predict via CFD the aerodynamic behavior of the model in the wind tunnel. Then using the same mesh definition, solver, turbulence model and setting adopted for free air prediction, the latter could be considered validated in a "transitive" way.

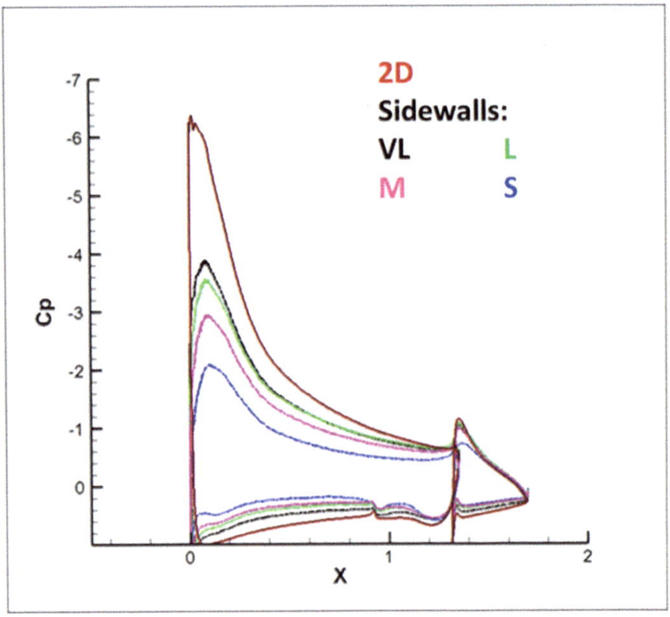

Fig. 31 The pressure coefficient distribution in the centerline of the model

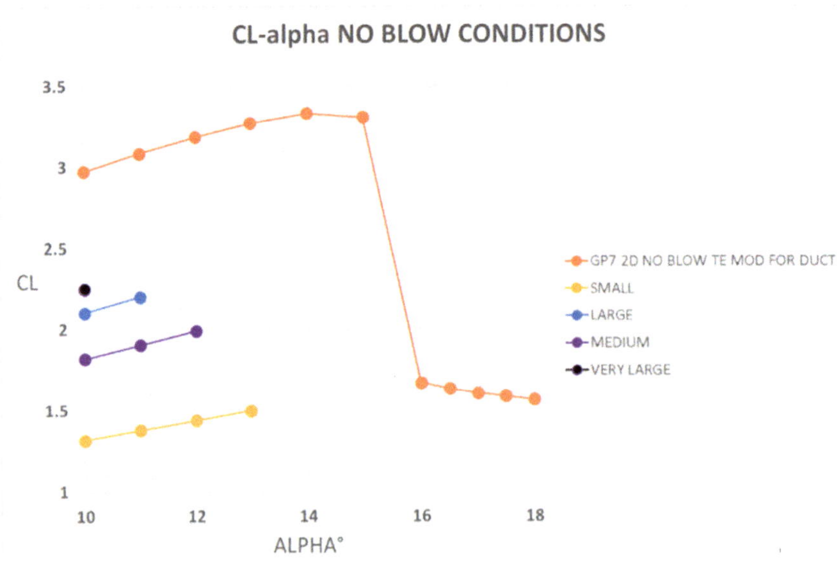

Fig. 32 The lift coefficient values increasing wind tunnel model sidewalls versus the 2D airfoil with modified trailing edge to house the blow duct

6 The Digital Twin Model

The digital twin CFD model originates from the CAD file kindly granted by the Von Karman Institute of Fluid Dynamics. The delivered geometry took in account the main details, including the side wooden fairings, the metal beams model supports, the floor shape and finally the test model (Fig. 33).

The CFD model considers the geometry half right only, Fig. 34.

The test model air-feeding model considered the last pipe part, Fig. 35.

6.1 Mesh and Simulation Setting

The computational domain extended to the wind tunnel hangar to simulate the undisturbed zero speed flow. This choice left the flow to spill out during the initial part of the computation, when the test model acted as strong obstacle for flow.

The wind tunnel mesh details included the upstream contraction and the downstream flow exit ducts, both with the proper edge lips. The mesh took in account the side fairing walls for the model rotation and force balance system, the rooftop wall, the model support frame and the model itself. As regard the model, the mesh considered the inner model air feeding as shown in Fig. 35.

The mesh (Fig. 36) counted $67e^6$ cells tetra and prism for boundary layer discretization, reaching the full available RAM.

The Metacomptech CFD++ [3–5, 7] was the used code, solving Preconditioned/ Compressible Pressure based equations. The k-ω SST [6] and k-ε realizable [8] turbulence models were both tested being the second model less computational time consuming. The increase in computational effort for k-ω SST is due to the implementation of blending functions that require the calculation of the distance from the

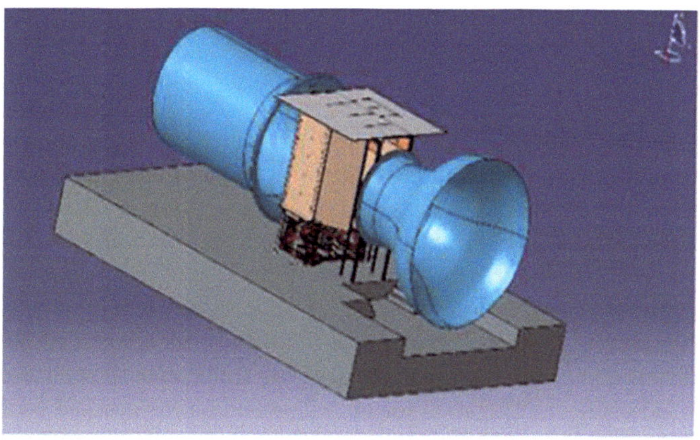

Fig. 33 The CAD file granted by the Von Karman Institute of Fluid Dynamics

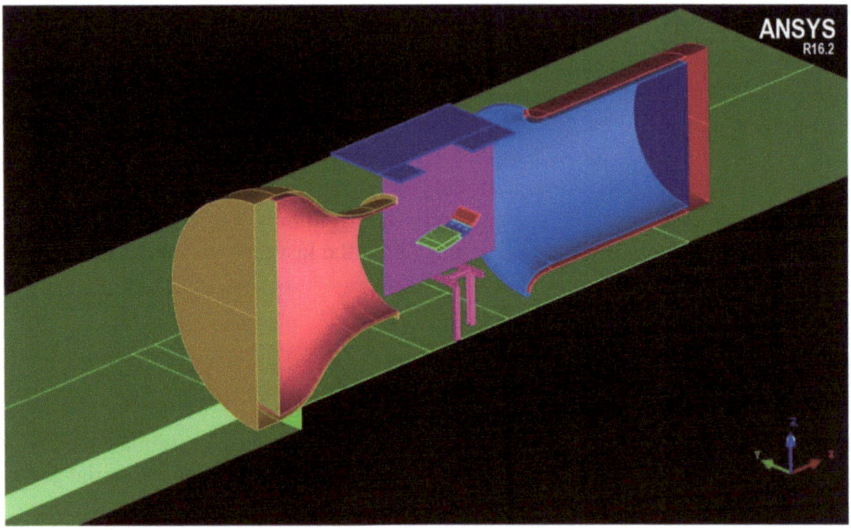

Fig. 34 The CAD part considered for the digital twin model

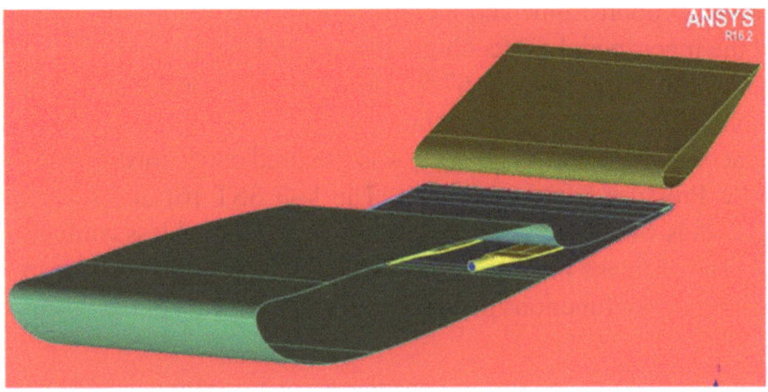

Fig. 35 The CAD part considered for the digital twin model—airfoil. Blow duct detail

wall of all the first centroids of the cells near the wall. The computation time occupied by this preliminary operation increases with the number of cells present in the computational domain and, in this case, has a time extension of 4 h.

Alternatively, the k-ε Realizable turbulence model was considered [8], which introduces a term into the transport equation of the rate of dissipation of turbulent kinetic energy (ε) to improve the response to adverse pressure gradients. k-ω was preferred, despite the higher computational time effort, for a better response to the adverse gradient compared to k-ε realizable, whose application showed an early stall.

The second order discretization for both flow and turbulence had been set. The desired wind tunnel speed (42.5 m/s) resulted from the application of Total Pressure and Temperature Reservoir boundary condition at the inlet of upstream duct and of

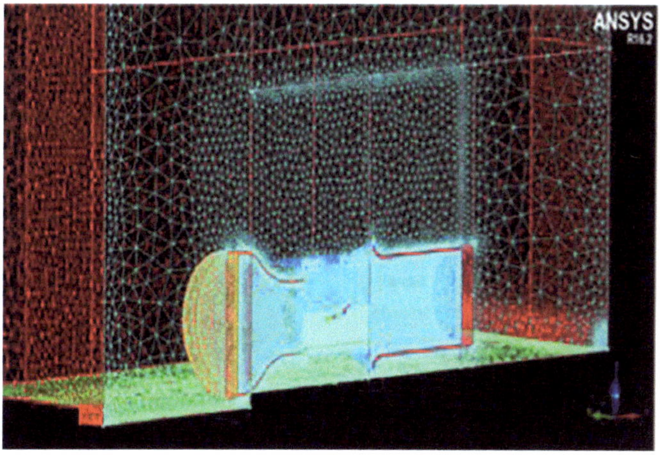

Fig. 36 Tetra-prism mesh

Outflow Back Pressure boundary condition at the outlet of downstream duct. The external box has been set as Characteristic Based Inflow-Outflow, which considers the undisturbed flow conditions as the virtual state external to the domain, for Riemann problem solution at the boundary domain. The Symmetry boundary condition had been applied to the geometry symmetry plane and the Adiabatic Viscous Wall boundary conditions to the wind tunnel and model walls.

The Total Pressure and Temperature Reservoir boundary condition applied at air inlet surface of the blow duct for each module (Fig. 35) set the prescribed Total Pressure Ratio.

6.2 Results

The first simulation considered the test chamber geometry initially proposed by the Von Karman Institute of Fluid Dynamics. The arrangement included the side wooden fairings and the top wall. The mesh considered the test model support beams. The test model configuration was no-blow, the flap at 30°, the Angle of Attack 8°.

The post processing indicated a bending up of the flow just downstream the test model with a flow Angle of Attack reduction and consequently loss of lift with respect to the predicted free stream simulation with infinite side walls (2D simulations), as showed in Fig. 37. A new mesh without the support beams, generated to check their impact on the flow, showed as they increased the flow bending up (Fig. 38).

The simulation indicates in addition the presence of some effect acting on the wind tunnel walls. A flow separation is clearly visible downstream the lower lip rear the test model (Fig. 39).

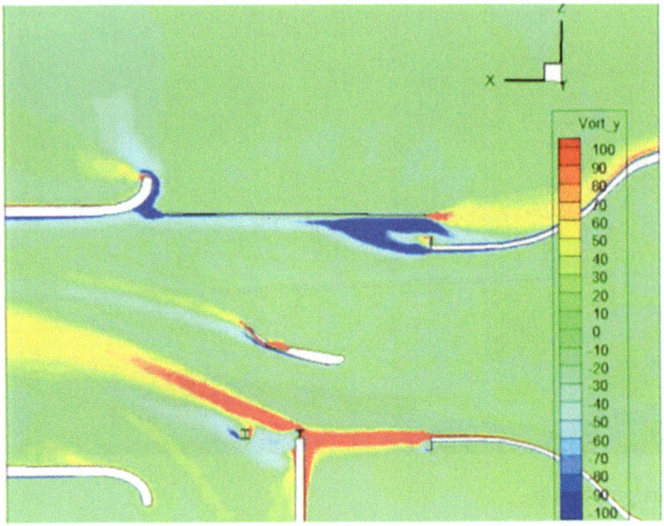

Fig. 37 Wind tunnel arrangement-airfoil aerodynamic interference

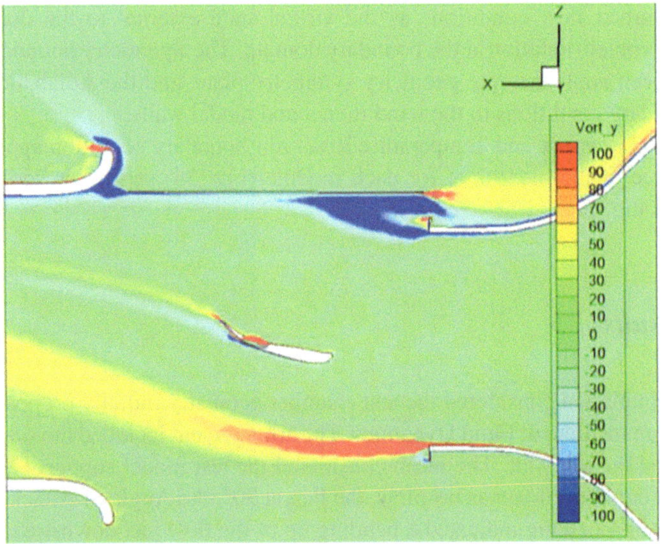

Fig. 38 Wind tunnel arrangement-airfoil aerodynamic interference

Following these simulations, the Von Karman Institute of Fluid Dynamics proposed to close the lower side with a plate to limit any flow interference with the support beams (Fig. 40) and a new simulation showed the reached correction (Fig. 41).

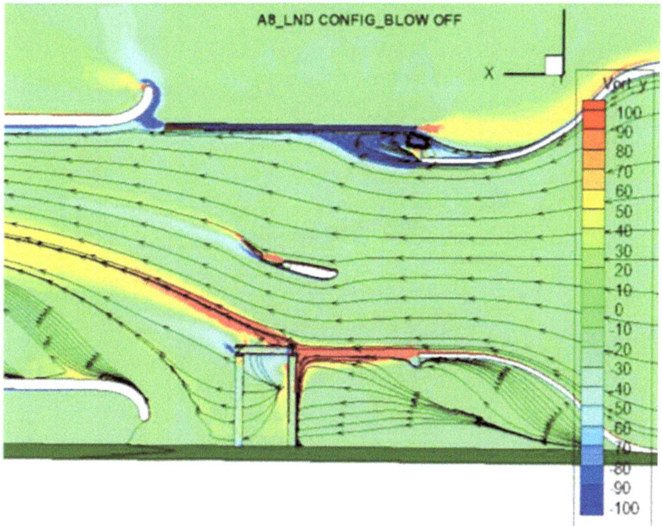

Fig. 39 Wind tunnel rear lower wall flow detachment

Fig. 40 Bottom closing plane

The flow was straighten up and the flow separation on the lower lips of the downstream duct disappeared.

Figure 42 shows the loop enlargement of the Pressure Coefficient due to the wind tunnel test chamber improvement.

The Von Karman Institute of Fluid Dynamics proposed a further improvement of the wind tunnel arrangement opening a window in the upper closing plate. The twin

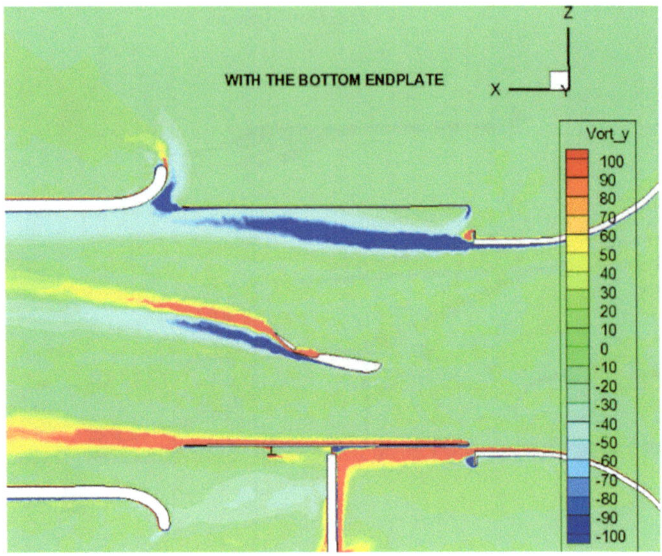

Fig. 41 Wind tunnel test chamber correction

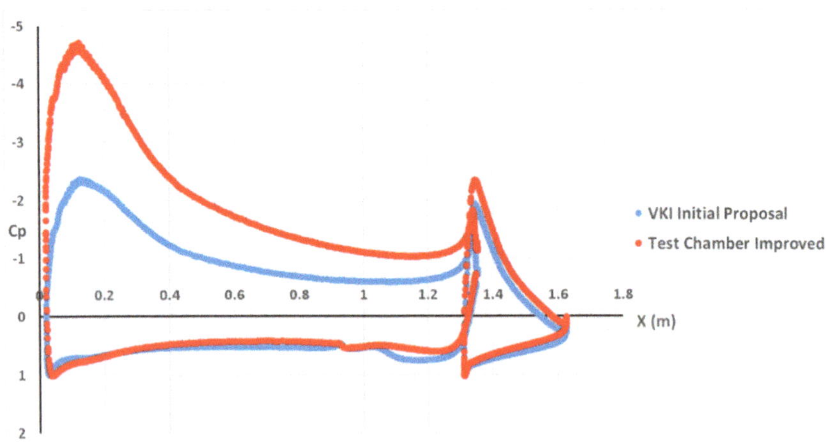

Fig. 42 Wind tunnel test chamber improvement

digital CFD simulation, in this case also allowed for an accurate prediction of the effect, showed in Fig. 43.

The upper closing plate opening reducing the wall interference, contributed to a further increasing of the test model Angle of Attack, as can be appreciated looking at the increased suctions on the airfoil upper surface.

Table 3 shows the comparison of the reached Lift Coefficient values resulting from the several test chamber modifications.

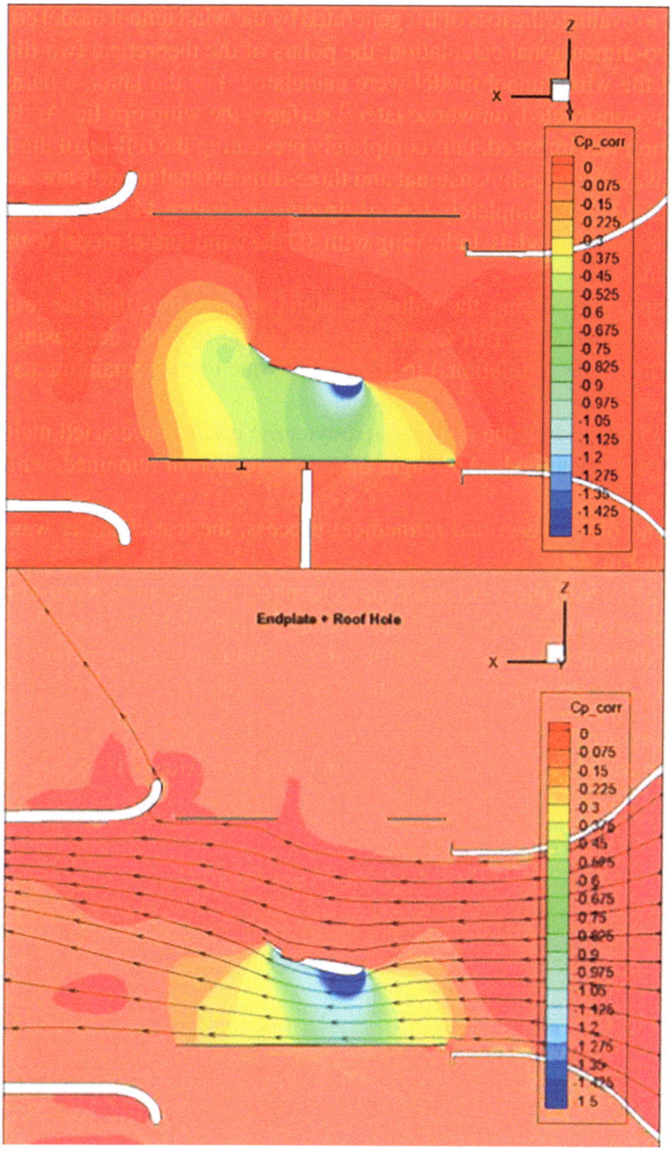

Fig. 43 Wind tunnel test chamber correction

Table 3 Lift coefficient after the wind tunnel test chamber improvements

Flap = 30°, alpha = 8°	C_L
No bottom plate, rooftop closed	1.87
Bottom plate, rooftop closed	2.63
Bottom plate and hole on rooftop	2.86

In order to evaluate the loss of lift generated by the wind tunnel model compared to the pure two-dimensional calculation, the polars of the theoretical two-dimensional profile and the wind tunnel model were calculated. For the latter, a parallelepiped far field was constructed, on whose lateral surfaces the wing tips lie. At these tips, a wall condition was imposed, thus completely preventing the roll-up of the flow at the tip. In this way, the two-dimensional and three-dimensional models are comparable, with the latter being completely free of tip effects. Figure 44 shows the lift curves obtained for the two models, indicating with 3D the wind tunnel model with infinitely extended side walls.

This graph, considering the values at alpha = 8° shows that the test chamber arrangement generated a Lift Coefficient loss equal to 0.68, decreasing from CL = 3.54 (2D and 3D simulations) to CL = 2.86, due to the remaining aerodynamic interferences.

Figure 45 shows that the residual aerodynamic interference acted mainly on the leading edge suction peak. An Angle of Attack reduction remained, which can be evaluated in 7°.

At the end of the described refinement process, the test chamber was frozen as presented in Fig. 46.

After the freeze of the wind tunnel test chamber configuration, some comparisons between the experimental data and the Digital Twin model have been done in terms of Lift Coefficient and Pressure Coefficient distribution over the airfoil.

Two very different configurations have been selected:

- Blow Pressure Ratio 1, flap at 30° and Angle of Attack 8°
- Blow Pressure Ratio 1.48, flap at 15° and Angle of Attack 4°.

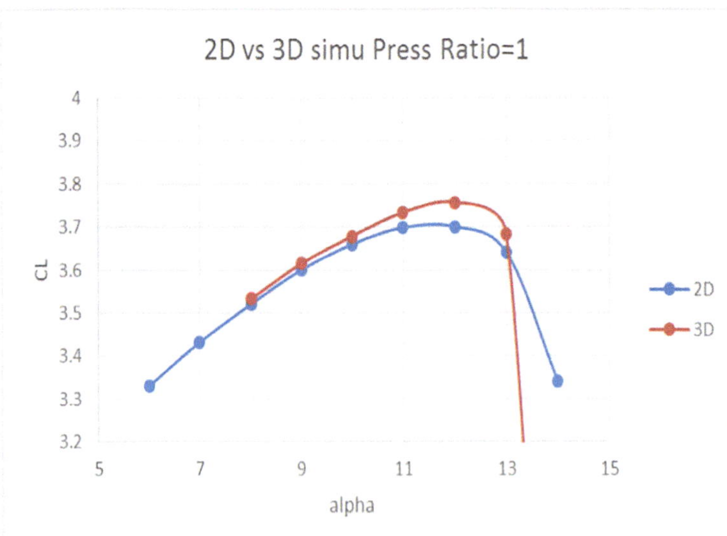

Fig. 44 Flap 30°, two and three dimensional (infinite end plate) simulations

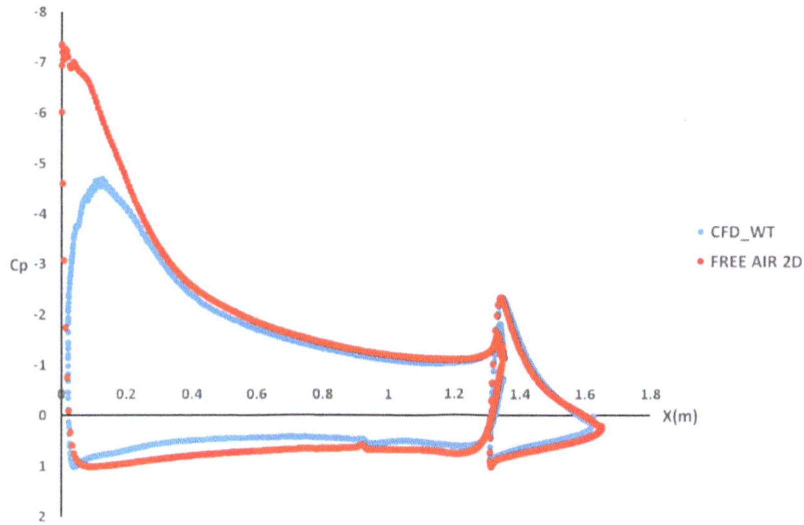

Fig. 45 Flap setting = 30°, V = 42.5 m/s, pressure ratio = 1 and angle of attack = 8°. 2D free air simulation versus wind tunnel digital twin CFD simulation

Figure 47 and Table 4 show the first configuration comparison:

Both the CFD Pressure Coefficient distribution and integral aerodynamic coefficients values are in good agreement with the corresponding experimental data. This result confirms the reliability of the Digital Twin model.

Figure 48 shows larger Pressure Coefficient distribution discrepancies between CFD and Experimental results. The gap is located mainly in the model main wing upper surface and its effect on lift coefficient is synthetized in Table 5.

The coarseness of the mesh in the zone of the blowing jet, due to the computer RAM limitations, was the main cause of the under-prediction of the Lift Coefficient with respect to the Wind Tunnel experiment. Figure 49 shows the impact of that coarseness on the definition velocity gradients of the blow jet thanks to the comparison with a pure 2D simulation. In the 3D case (Fig. 49 right side), the jet results smeared, weakening the airfoil circulation.

7 Conclusions

The presented activity showed the important role of the Digital Twin approach in evaluating the quality of the experimental framework. The simulation of the actual wind tunnel test arrangement, considering its hangar, allowed for a preliminary wind tunnel flow analysis that highlighted important criticalities, helping in prevent time consuming and expensive correction loops.

Fig. 46 Wind tunnel test chamber final arrangement

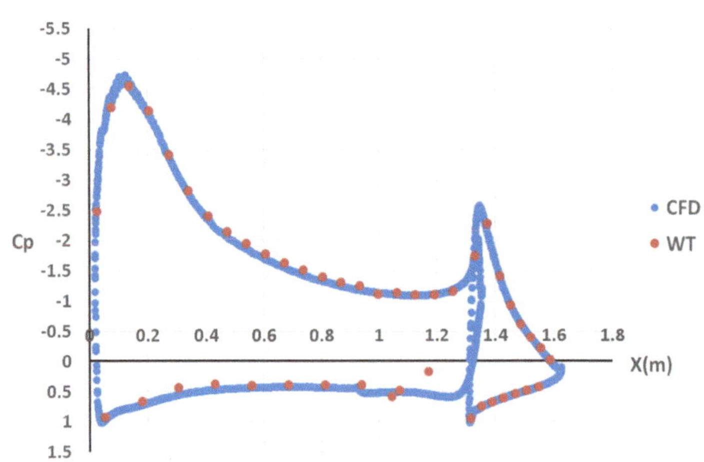

Fig. 47 Flap setting = 30°, V = 42.5 m/s, pressure ratio = 1 and angle of attack = 8°. CFD and experimental pressure coefficient distribution

Table 4 Flap setting = 30°, V = 42.5 m/s, pressure ratio = 1 and angle of attack = 4° and 8°

	CFD		Experimental		Δ CFD versus experimental [%]	
Angle of attack [°]	CL	Cm	CL	Cm	ΔCL	ΔCm
4	2.680	– 0.75	2.626	– 0.70	+ 2	+ 6.3
8	2.860	– 0.625	2.892	– 0.66	– 1.2	– 4.3

CFD and experimental integral aerodynamic coefficients comparison

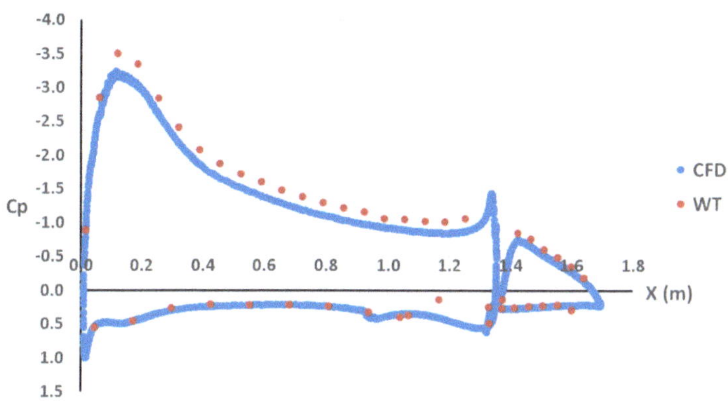

Fig. 48 CFD and experimental pressure coefficient distribution. Flap setting = 15°, V = 42.5 m/s, pressure ratio = 1.48, angle of attack = 4°

Table 5 Flap setting =15°, V = 42.5 m/s, V = 42.5 m/s, pressure ratio = 1.48, angle of attack = 4°

Flap 15°, Angle of attack 4°, P ratio 1.48	
CL experimental	2.25
CL CFD	2.03
Δ CFD versus experimental [%]	– 10

CFD and experimental integral aerodynamic coefficients comparison

The CFD simulations suggested the introduction of several test chamber modifications to correct flow behavior or confirmed the effectiveness of proposed architecture solutions.

The very large test model dimensions with respect to the wind tunnel test chamber led to introduce a transitive validation concept to overcome the differently unsurmountable wall interference and flow blockage effects. The Digital Twin model

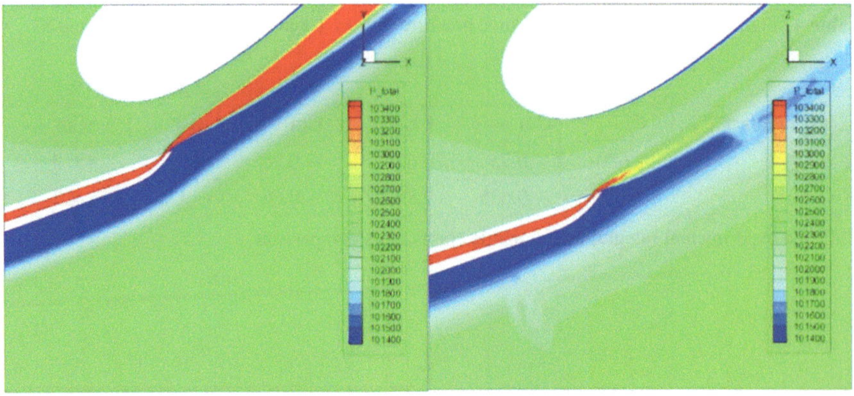

Fig. 49 Mesh refinement comparison. Left side: 2D very fine mesh. Right side: 3D coarse mesh

served to demonstrate the reliability of the mesh generation settings, solver and turbulent model selection making more trustable the free air simulation results.

Acknowledgments The project MOTHIF (MOdel Testing of HIgh LiFt system) has received funding from the Clean Sky 2 Joint Undertaking (JU) under grant agreement No. 865267. The JU receives support from the European Union's Horizon 2020 research and innovation programme and the Clean Sky 2 JU members other than the Union.

Disclaimer The present paper reflects only the author's view and JU is not responsible for any use that may be made of the information it contains.

Data Availability Statement Data that support the findings of this study are available from the corresponding author upon reasonable request.

References

1. Jones, G.S., Joslin, R.D.: Proceeding of the 2004 NASA/ONR Circulation Control Workshop. NASA/CP-2005-213509 (2005)
2. Massachussets Institute of Technology homepage, web.mit.edu/drela/Public/web/xfoil
3. Peroomian O., Chackravarthy S., Goldberg U.C.: A 'grid-transparent' methodology for CFD. AIAA-97-0724 (1997)
4. Wilcox, D.C.: Turbulence Modeling for CFD. DCW Industries, Inc., La Canada, California (1998)
5. Menter, F.R.: Two-equation Eddy-viscosity turbulence models for engineering applications. AIAA J. **32**(8), 1598–1605 (1994)
6. Palaniswamy, S., Chackravarthy, S.R.: Stability of explicit schemes in the physical and frequency domains. J. Comput. Phys. **124**, 162–176. Article no. CP965632 (1997)
7. Chackravarthy S.R., Szema K.-Y., Goldberg U.C., Gorski J.J., Osher S.: Application of a new class of high accuracy TVD schemes. AIAA85-0165 (1985)

8. Shih, T.-H., Liou, W.W., Shabbir, A., Yang, Z., Zhu, J.: A new k-epsilon Eddy-viscosity model for high Reynolds number turbulent flows—model development and validation. Comput. Fluids **24**(3), 227–238 (1995)
9. Harten, A., Engquist, B., Osher, S., Chackravarthy, S.: Uniformly high order accurate essentially non-oscillatory schemes, III. J. Comput. Phys. **131**, 3–47. Article no. 0051 (1996)

The Role of High-Fidelity Computer-Aided Engineering (CAE) Multi-physics Design Within the Research Institutes of the National Institute for Astrophysics (INAF)

G. Lombardi, E. Di Meo, A. Lopez, S. Fabiani, M. Feroci, and M. E. Biancolini

Abstract In the domain of aerospace engineering, the design of scientific payload instruments for spaceborne applications presents a significant challenge due to the inherent complexity arising from multi-physics phenomena and the associated modeling requirements. Simulating these interdependent physical processes necessitates the implementation of intricate numerical models, often leading to computationally expensive calculations across disparate temporal and spatial scales. Integration of these multi-physics models is the key to understand how the different physics processes affect each other. Furthermore, only optimal efficiency of these coupled physic models enables whole domain simulations to validate the models and integrate them in the tools. In this work we describe the role of the high fidelity computer-aided engineering (CAE) multi-physics in the National Institute of Astrophysics (Institute of Space Astrophysics and Planetology (IAPS)/Italian National Astrophysics Institute (INAF)), the design solutions adopted for the most important thermo-mechanical design drivers for two different kind of space missions: the Large Area Detector (LAD) module that will mount on the enhanced X-ray Timing and Polarimetry (eXTP) mission, a major project of the Chinese Academy of Sciences (CAS) and China National Space Administration (CNSA) currently performing a phase B study and proposed for launch in 2028, and the payload of CUbesat Solar Polarimeter (CUSP), a project of the Institute, funded by the Italian Space Agency (ASI) for a phase A study that consists in a constellation of two CubeSats 6U orbiting around the Earth to measure the linear polarization of X-rays of solar flares in order to improve the knowledge of these violent phenomena.

Keywords Mechanical design · Optimization · FEM analysis · Multi-physics · Space

G. Lombardi (✉) · S. Fabiani · M. Feroci
IAPS-INAF—National Institute of Astrophysics, Rome, Italy
e-mail: giovanni.lombardi@iaps.inaf.it

G. Lombardi · E. Di Meo · A. Lopez · M. E. Biancolini
Department of Enterprise Engineering, University of Rome "Tor Vergata", Rome, Italy

M. E. Biancolini et al. (eds.), *Digital Twins for Aerospace*, Springer Aerospace Technology, https://doi.org/10.1007/978-3-032-05022-9_9

1 Introduction

According to Neelamkavil, "a model is a simplified representation of a system intended to enhance our ability to understand, predict and possibly control the behaviour of the system" [1]. Advances in numerical computation have empowered engineers and researchers to leverage simulation techniques for comprehensive multi-physics analysis of aerospace instrument performance. Throughout a scientific project's life cycle, mathematical models are constructed based on fundamental physical principles. These models serve as a cornerstone for knowledge extraction, guiding optimal decision-making across all development phases.

To ensure accurate design representation and compliance with space mission requirements and the operational environment of the spacecraft, a multi-physics approach is paramount. This methodology surpasses traditional methods by enabling the coupling of any number of relevant physical phenomena. Consequently, depending on the specific application, diverse, interconnected models can be employed to comprehensively describe the mission's behavior.

The goal of this chapter is to explain the high-fidelity cae solutions adopted in some spacecraft instrument designed in the National Institute of Astrophysics (IAPS-INAF) in Rome. Simulation models have a vital role and can be used by engineers to make decisions at all phases of a space program: the conceptual and design phase, control design testing, stability analysis, system design and envelope expansion. The proposed models include thermo-mechanical analysis, design optimization, morphing skills and a cross-platform tool for the mechanical Computer Aided Design (CAD) implementation in a physics simulator software. The results show that the simulation models presented in this paper are accurate and could be further used to analyse other space missions and for the future multi-physical challenges for space science and exploration.

2 Background and Method

This paper explores various space mission typologies. For each, it describes the preliminary design and the multi-physics numerical analysis solutions adopted for the critical thermo-mechanical drivers based on mission requirements.

Specifically, the paper details the mechanical design of a single module within the LAD instrument for the Sino-European eXTP space mission. It covers all subsystems, the initial results of static and dynamic finite element analysis, and a thermo-elastic analysis of the LAD module. This analysis investigates how tension, deformation, and structural displacement change under extreme thermal conditions [6]. The goal is to optimize the collimator frame design by minimizing the misalignment angle between the collimator and the forward direction, shown in Fig. 1.

Furthermore, the paper examines the mechanical design of the payload for a CubeSat mission (CUSP), focusing on the truss interface between the payload and

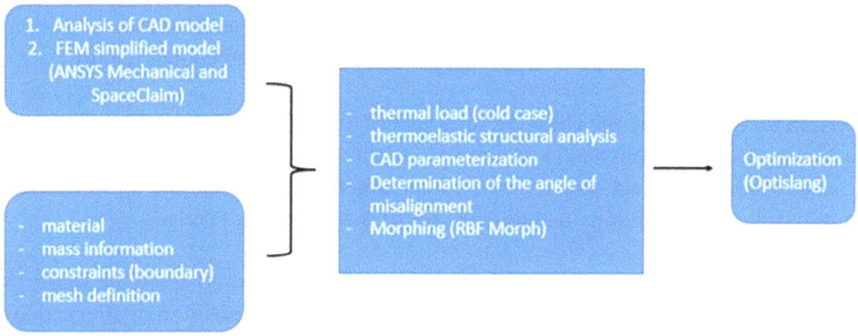

Fig. 1 Collimator frame optimization

the platform. It also presents the development of a tool enabling interaction between a physics Monte Carlo simulator (GEANT4) and the CAD file. Key geometric features were parameterized [7], and an optimized design is proposed. This optimization [11] aims to understand how the main parameters of merit respond to variations in geometric characteristic combinations, while considering the geometric constraints imposed on the volume and mass of the structures.

2.1 Software

The software used for this project are:

- SolidWorks 2022
  SolidWorks [13] is a three-dimensional parametric design software, produced and marketed by Dassault Systèmes. Useful for the design of mechanical devices, even complex ones, it involves the creation of 2D and 3D drawings of solids and surfaces, through a parametric geometric system. Parametric design allows to set up numerous types of relationships (parallelism, concentricity, perpendicularity, collinearity, equality, coaxiality, symmetry and many others).
- Systema THERMICA
  The Thermica suite [15] provides a complete package to perform the thermal design. It offers modules to compute radiative couplings, conductive couplings, external fluxes (e.g. solar, planetary) gathered in the software with interactive toolboxes to build the geometry, define a trajectory or a kinematics, specify an analysis scenario, visualize the results, set the analysis process and post-process the results.
- ANSYS SpaceClaim
  ANSYS Space Claim is a parametric solid modeling software available in the ANSYS package that operates mainly according to the direct modeling logic,

thanks to which it is possible to define the construction and modification operations of the desired geometry, without the need to define interdependent features.

- ANSYS Meshing
  ANSYS Meshing is the software responsible for generating the calculation grid. This process is automated in its simplest application although it is possible to add control settings to act directly on those parameters that define the mesh in terms of morphology, topology and dimensions of the constituent cells.
- RBF Morph
  Radial basis functions (RBF) based mesh morphing allows to adapt the shape [8, 9] of a computational grid onto a new one by updating the position of all its nodes. Usually nodes on surfaces are used as sources to define the interpolation field that is propagated into the volume mesh by the rbf. The method [10] comes with two distinctive advantages that makes it very flexible: it is mesh independent and it allows a node wise precision. There are however two major drawbacks: large data set management and excessive distortion of the morphed mesh that may occur.
- ANSYS Mechanical
  ANSYS Mechanical is the finite element calculation software that allows to set up and carry out structural analysis.
- ANSYS DesignXplorer
  ANSYS DesignXplorer is a software tool that empowers users to generate, explore, and query within a defined parametric domain. In this project, it was instrumental in selecting the mesh during the initial reconstruction phase of the simplified finite element model (FEM) for the LAD module. Subsequently, it played a crucial role in various structural optimizations. DesignXplorer's built-in algorithms operate by first defining a multidimensional domain. Each dimension represents an input parameter previously defined within the model. The software then calculates the desired output parameters, establishing a functional relationship between them and the entire input domain. In this project, the Response Surface Optimization tool within DesignXplorer was specifically leveraged for mesh convergence and optimization.
- OptiSLang
  OptiSLang [14] is a software platform for cae based sensitivity analysis, multidisciplinary optimization (MDO) and robustness evaluation. It provides a framework for numerical Robust Design Optimization (RDO) and stochastic analysis by identifying variables which contribute most to a predefined optimization goal. This includes also the evaluation of robustness, i.e. the sensitivity towards scatter of design variables or random fluctuations of parameters. This procedure can also be used for a pre-optimization before global optimizer (evolutionary algorithms, Adaptive Response Surface Methods, Gradient-based methods, biological-based methods) are used for a direct single-objective optimization.
- ANSYS Workbench
  The Workbench platform is the tool designed to put in communication the various modeling and numerical analysis software that fall within the ANSYS landscape, facilitating the management of the project that is articulated there through various work steps. Its key feature is to allow the transfer and sharing of input and output

data not only between the different tools used in pre and post processing, but also between multiple simulation models (which allows for the easy implementation of sophisticated multi-physical simulations).

The initial geometry models were created using Solidworks 2022 software. However, the shapes were subsequently reconstructed and simplified within ANSYS SpaceClaim. This software's standard features facilitated this process. Furthermore, leveraging the advanced capabilities of SpaceClaim, a procedure was implemented to exploit symmetry and conserve structural constraints, even after model lightweighting.

3 eXTP—Enhanced X-Ray Timing and Polarimetry Mission

The eXTP [2] mission is a high-energy astrophysics space science mission designed to study fundamental physics under extreme conditions of density, gravity and magnetism. The mission aims at determining the equation of state of matter at supranuclear density, understanding the dynamics of matter in strong-field gravity and measuring effects of quantum electro dynamics (QED). In addition to investigate fundamental physics, eXTP is going to be a very powerful observatory for X-ray astrophysics that will provide observations of unprecedented quality on a variety of galactic and extra galactic objects.

The scientific payload of the eXTP satellite consists of four scientific instruments: the spectroscopic focusing array (SFA), a set of 9 telescopes equipped with Silicon Drift Detectors, the polarimetry focusing array (PFA), a set of 4 telescopes equipped with the polarization-sensitive Gas Pixel Detectors, reaching an effective area 5 times larger than the polarimeter onboard the Imaging X-ray Polarimetry Explorer (IXPE) mission [5]), the wide field monitor (WFM) [4], a coded mask instrument simultaneously imaging $\sim$1/3 of the sky, with arcmin resolution) and the LAD, subject of this paper.

In Fig. 2 its shown its current configuration, with the allocation of the four scientific instruments and the sun shield.

3.1 LAD Instrument

The Large Area Detector [3] (LAD) is a 3 m^2-class instrument with $\sim$ 10 times the collecting area of the largest past timing missions, which holds the capability to revolutionize the studies of X-ray variability from X-ray sources on the millisecond time scales.

The LAD payload is a large array of X-ray detectors, 640 silicon drift detector (SDD), electrically and mechanically organized in groups of 16, referred to as

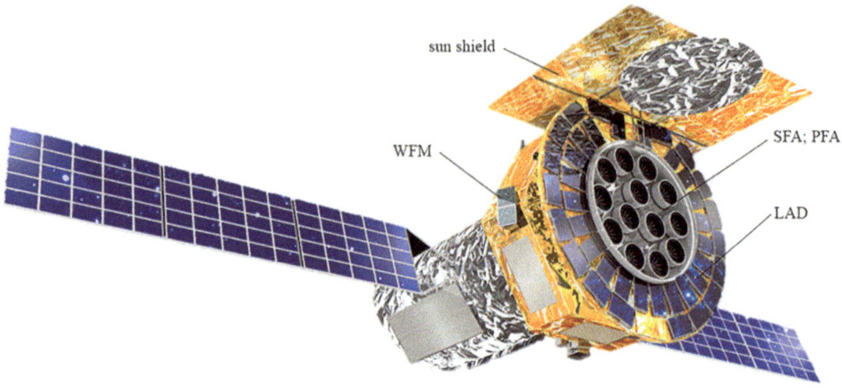

Fig. 2 eXTP satellite concept

modules, for a total of 40 modules. Ten modules are electrically connected to the same panel back end electronics (PBEE), the red's square, and each of the 4 PBEE is connected to the instrument control unit (ICU), the blue's square, nominal and redundant, ICU is then connected to the spacecraft (S/C).

The current baseline for the LAD payload configuration is based on 40 modules in a double circumferential arrangement on an annular optical bench.

The organization and structure of the LAD is shown in Fig. 3.

The study of the energy-resolved timing properties of the X-ray emission of cosmic sources requires the accurate measurement of the time of arrival (TOA) and energy of the largest number of photons from the target source. The unambiguous identification of the target source in this type of experiment is most effectively achieved by narrowing the field of view by means of an aperture collimator, down to a level (typically $\leq \sim 1°$) large enough to allow for pointing uncertainties yet small enough to reduce the aperture background (cosmic diffuse X-ray background) and the risk of source confusion.

The LAD is therefore designed as a classical collimated experiment.

The requirement to acquire X-ray events with high time and spectral resolution and high statistical significance (to study astrophysical sources in the time/frequency/energy domain) has led to the LAD instrument described here, made possible by developments in large-area silicon detectors and micro-collimator structures.

The primary features are:

- A large effective area for a single on-axis source;
- No imaging, but instead collimation to maximize on-axis effective area;
- High energy resolution.

The field of view of the LAD will be limited to $< 1°$ by X-ray collimators. These are developed by using the technique of micro-capillary plates, the same used for

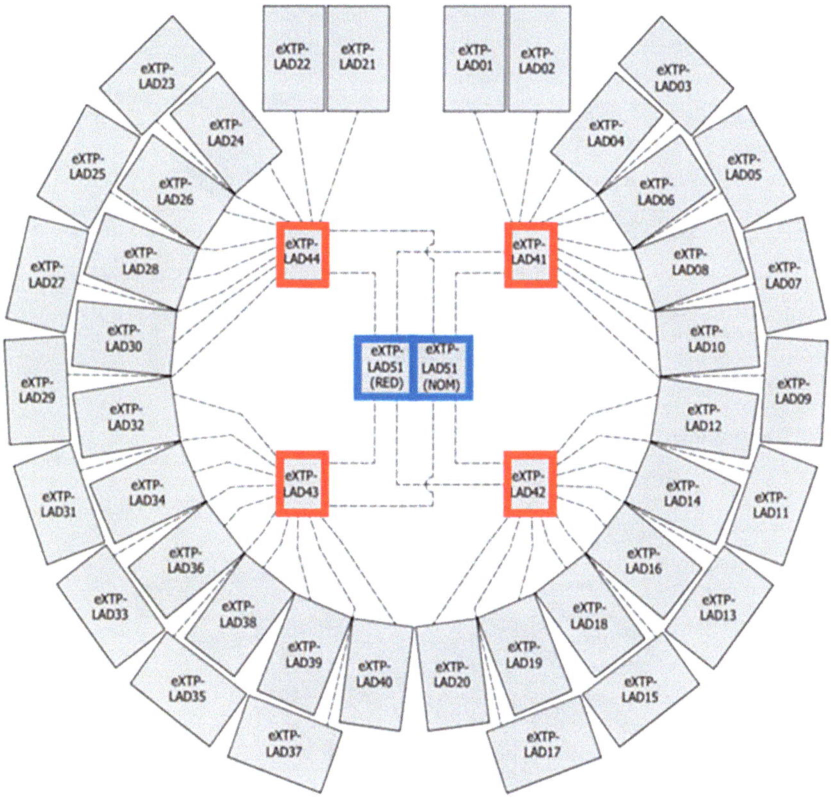

Fig. 3 LAD instrument schematic

the micro channel plates: a ~5 mm thick sheet of Lead glass is perforated by a huge number of micro-pores, ~83 μm diameter, ~16 μm wall thickness.

3.2 Mechanical Design

Figure 4 shown the module, that contains:

- An aluminium alloy Collimator Frame with 16 co-aligned MPO tiles (in positions corresponding to the 16 SDD), clamped to the frame;
- Underneath each collimator tile MPO and the detector is a thin (~ 1 μm) Kapton film, coated with ~100 nm aluminium on each side, is hosted;
- A detector frame containing 16 SDD+ front end electronics (FEE);

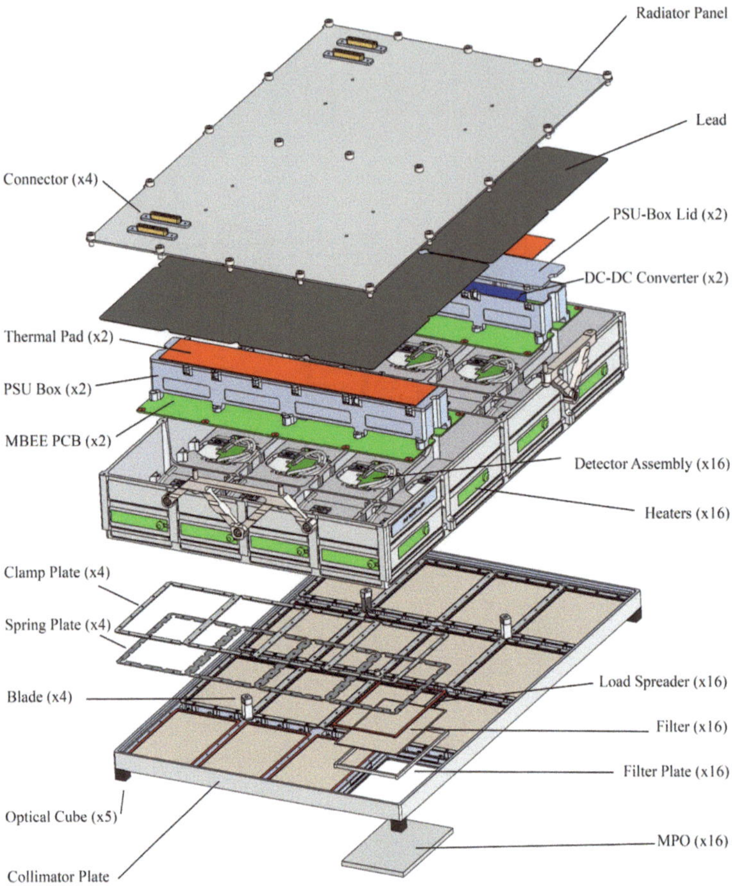

Fig. 4 Module exploded view, main components and subsystems

- module back end electronics (MBEE). Controls the ASICs and high voltage (HV) power supply unit (PSU) and reads out the digitized events. The MBEE is hosted in the same mechanical and electrical framework of the Detector Frame;
- HV/MV/LV PSU: This is the power supply for the SDD; the PSU is hosted in the same mechanical and electrical framework of the Detector Frame;
- A 300 μm lead back-shield, to reduce the background events in the SDD;
- A radiator to dissipate heat from the module (lower SDD temperature improves the energy resolution).

An heater belt is placed around the module. The heating process during annealing, which heat up the model up to 50 °C alters the silicon crystal structure, recovering the damages induced by particle radiation and balance the thermal gradient, following the mission's requisite, generated during the satellite orbits.

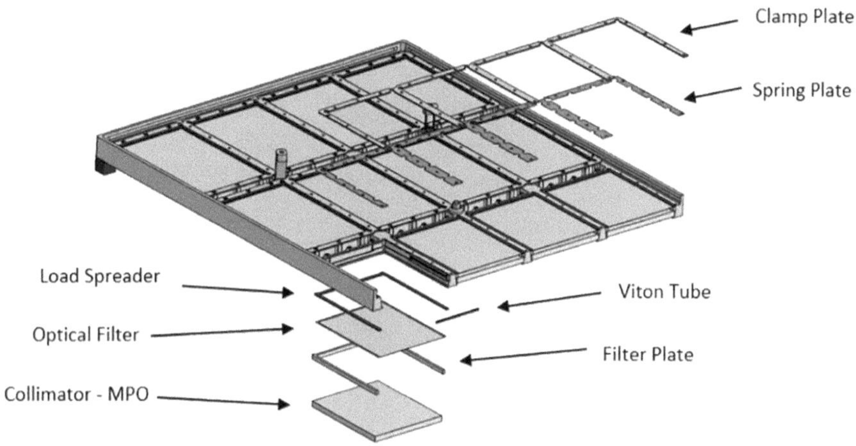

Fig. 5 Exploded view of collimator tray

Collimator Tray In the upper part of the collimator frame, Fig. 5 six optical cubes will be positioned and mounted to ensure that the stability and repeatability of the alignment are ≤ 10 arcseconds. The collimator tray hosts 16 individual MPO tiles, sheets of lead glass (Pb content about 40%) machined through the same process used to produce microchannel plates, in positions corresponding to each individual SDD detector. The MPO has a size of $111.0 \times 72.5 \times 5\,\text{mm}^3$ and fits into a recess in the aluminium alloy collimator frame.

A narrow portal frame pressure plate (load spreader) is placed on top of each MPO; also four clamp plates are positioned on top of the four spring plates. The spring guarantees sufficient preload, adequately spread on the MPO tile to prevent it from moving out of plane during vibration testing. The lateral movement of the MPO tiles is limited by the springs as well by four viton tubes, that should keep mechanical and chemical resistant properties at higher heat applications, placed at the collimator edges.

The collimator assembly is attached to the lower detector module assembly through four blades, slat fixing system, that are screwed into the lower detector module frame and pass through the collimator frame as it is lowered in place, this allows some movements between the frame assemblies.

The purpose of these blades is to reduce the thermo-elastic deformation of the collimator frame, rigidity is instead guaranteed by a central stud that, also, pass through the collimator frame.

An optical filter is used to complete the thermal and optical design, mounted on the collimator frame, above each detector, between the detector and the MPO at 3.5 mm distance to allow for any deformation at launch.

Detector Tray The fundamental LAD detection element is the Detector, composed of SDD+FEE (dubbed "Detector Assembly"—DA) and Collimator. The eXTP detector

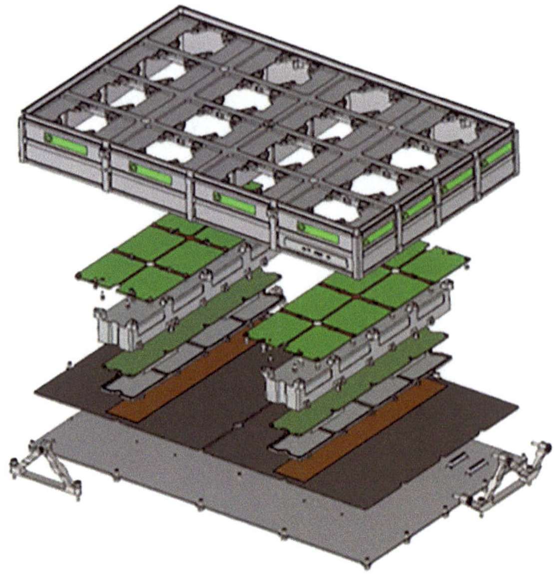

Fig. 6 Exploded view of detector tray

module assembly comprises sixteen Silicon Drift Detectors (SDD's) arranged in a four-by-four pattern.

The detectors are mounted in a monolithic aluminium alloy, Fig. 6, frame that supports each detector unit on four points with sufficient compliance to cater for thermal expansion variation between the frame and the detector FEE PCB. For each detector there is a conductive link via the structure to the radiator as shown in Fig. 7.

The FEE size is as close as possible to the SDD (120×72 mm) in order to maintain a sensitive area of the module close to 100% when all FEE are mounted into the module.

Two module back-end electronics (MBEE) PCB assemblies with HV supplies attached are mounted behind the bank of detectors, each BEE/HV assembly interfacing and serving eight detectors.

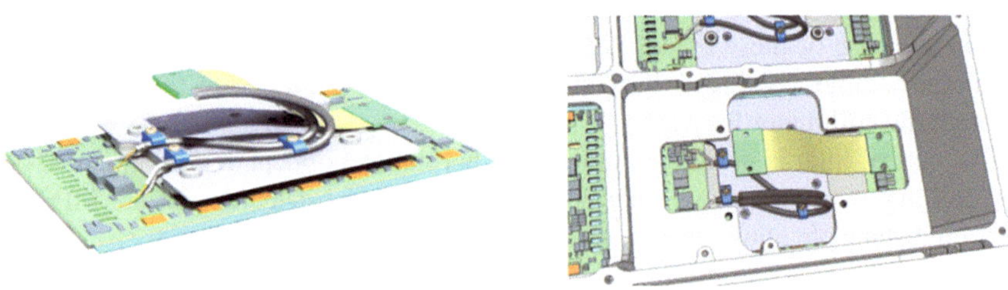

Fig. 7 Silicon drift detector with front end electronics

A lead shield and an aluminum alloy radiator panel screwed directly to the back of the detector module body then completes the build of the lower detector module assembly.

The radiator is a single aluminum sheet of 2 mm thick, this thickness is adopted to minimize the thermal gradient across this radiator and therefore optimizes the detector temperature over each orbit.

Each module will be mounted kinematically in the payload structure to mechanically isolate each module from any deformation of the interface.

In the end, the interface between the modules and the spacecraft is guaranteed by applying three bipods, two of which are standard in the long-side and one W-bipod on the short side of the module.

3.3 Numerical Analysis and Optimization

Model setup The first step in preliminary finite element analysis (FEA) is the creation of a 3D CAD model. This can be done using various software programs. In this case, SolidWorks 2022 was used. However, geometries can also be imported into commercial FEA software from most major modeling programs.

Once the detailed model is created, it may be necessary to simplify it to avoid complex geometries (such as sharp edges and holes) that can lead to calculation difficulties. This can be achieved using a dedicated workbench within the commercial software, like SpaceClaim, which allows for fast and accurate modeling of the STEP file format. Before starting the modeling process, it's crucial to define a reference system and the units of measurement that will be used throughout the analysis.

- Material properties
  In order to start a simulation, each component of the model needs to be assigned a material with its corresponding mechanical properties. These materials, illustrated in Table 1, can be selected from the ANSYS library or created by manually defining their properties.
- Mass Information
  To account for elements not explicitly represented in the finite element method (FEM) such as fasteners, cables, and electronic components, an additional non-structural mass is added. The larger components are modeled as point masses with

Table 1 Material properties

Description	Density (kg m^{-3})	Young modulus (N m^{-2})	Poisson ratio
7075 T6 aluminum alloy	2796	7.1E+09	0.33
Titanium alloy	4620	9.6E+09	0.36
PCB laminate FR-4	1900	2.46E+09	0.136

Table 2 Mesh node and elements

Nodes	Elements (tetrahedral)
539,550	272,893

rigid connections to the printed circuit board (PCB). Meanwhile, the remaining electronic components, fasteners, and cables are represented as distributed mass divided into collimator frame (CF) and detector frame (DF). This approach helps to incorporate their contribution to the overall mass budget of the model.

- Interface Nodes
  After assigned the material to each body its necessary to apply the boundary conditions according the actual working conditions i.e. where it fixed and where it is free. The interface between satellite's truss and module is guarantee by the surface of bipods, illustrated in ??.
- Mesh generation
  this initial exploratory simulation campaign, the mesh was generated without any specific constraints. The results for the 3D model body and the 2D shells (printed circuit boards or PCBs) are presented in Table 2.

Collimator frame optimization This paragraph focuses on a specific module housed within the LAD instrument aboard the Sino-European eXTP satellite. Particular attention is given to the collimator frame, its bipods, and the mission requirements related to these components, such as alignment and natural frequency.

LAD modules operate under variable thermal conditions throughout the satellite's orbital cycle. These fluctuating temperatures induce varying degrees of deformation in the different components. The primary objective of this initial optimization is to analyze the thermal deformations experienced by the collimator frame, particularly in relation to its critical function of housing and precisely orienting the 16 collimator detectors. Subsequently, the design of the collimator frame will be improved by optimizing its ribs to achieve better pointing accuracy.

To achieve this objective, the study was conducted through the following steps:

- preliminary design of the module in a complete cad model meeting the requirements of the mission and export of the model in ANSYS SpaceClaim obtaining a simplified defeatured [12] model but compliant in terms of mass, volume and critical part of collimator frame object of this study;
- import into the simplified FEM model of the thermal load on components and interfaces associated with the most severe case (Cold Case);
- Definition of the parameters necessary to control the misalignment of the collimator with respect to the undeformed configuration of the collimator frame;
- Re-processing and parameterization of the cad geometry of the Collimator Frame through the rbf morph tool;
- Structural optimization of the Collimator Frame through Response Surface Optimization and OptiSLang.

Table 3 Temperature profile of cold case

I/F bipod-truss	+ 20 °C
I/F bipod-DF	− 48 °C
CF short side	− 67.3 °C
Detector frame	− 52 °C
Blade	− 61 °C
Central bolt	− 55.5 °C

Leveraging the symmetry of the module, this optimization process focused on just one half of the structure, effectively treating both halves as identical. This approach significantly reduced the computational cost of the simulation campaign. Using SpaceClaim, reinforcement ribs were added to the top of the collimator, resulting in a slight increase in its overall volume. The upper surface of the rib was then partitioned to create distinct surfaces that would serve as input for the optimization algorithm's mesh morphing capabilities. This step was crucial for facilitating the improvement and optimization of the collimator frame.

Finally, all non-functional elements supporting the collimator's connection to the detector frame were removed from the model.

Due to the varying thermo-physical properties of the materials, the components experience deformations during thermal cycles, leading to misalignment of the 16 MPO. Since the structure is not free to deform entirely due to internal and external constraints, these deformations induce internal stresses within the components.

To assess these deformations and verify the structural integrity, the temperatures corresponding to the most critical thermal case (the cold case) were imported into the finite element model. This information can be found in the accompanying Table 3.

The thermo-elastic analysis revealed that temperature variations from the reference temperature cause misalignment of the panels and overall deformation of the internal components.

Specifically, the study demonstrates that the collimator frame undergoes an upward inflection. Each point experiences an almost symmetrical displacement field along the x-axis, with larger displacements observed on the shorter side not constrained by the bipods. Conversely, the displacement field, shown in Fig. 8 along the y-axis is symmetrical with respect to the center of gravity of the collimator frame itself:

The observed displacement field indicates that the misalignment of the collimator frame is more pronounced at the outermost windows. This finding is likely due to the reduced stiffness and support in these regions compared to the areas constrained by the bipods.

Misalignment evaluation To analyze the displacement field in detail, the collimator frame was virtually divided into quadrants. The internal longitudinal sides of each

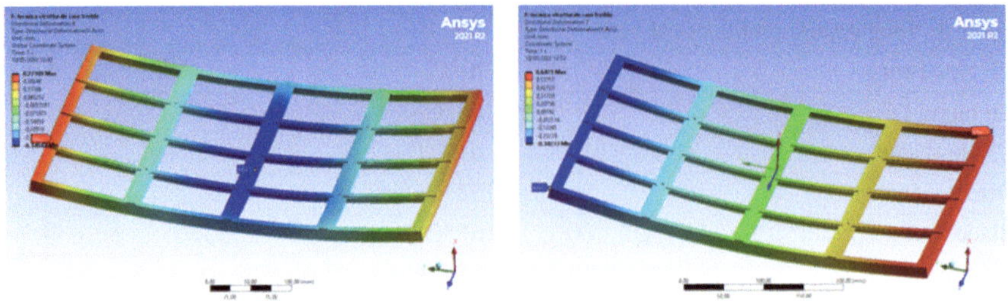

Fig. 8 Deformation along x-axis and y-axis

quadrant were then numbered following a matrix logic scheme in Fig. 9. This numbering system facilitates the precise identification and comparison of displacement values at specific locations on the collimator frame. Also, the following relation in Fig. 10 was used for the calculation of each single angle, expressed as a function of the displacements along all directions: To quantify the overall misalignment trend

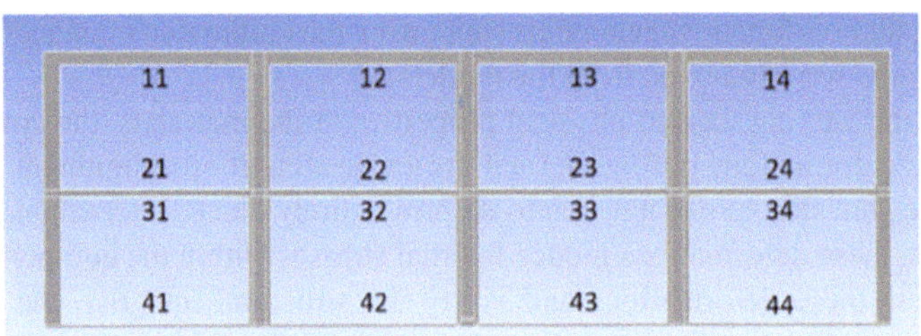

Fig. 9 Subdivisions of collimator frame

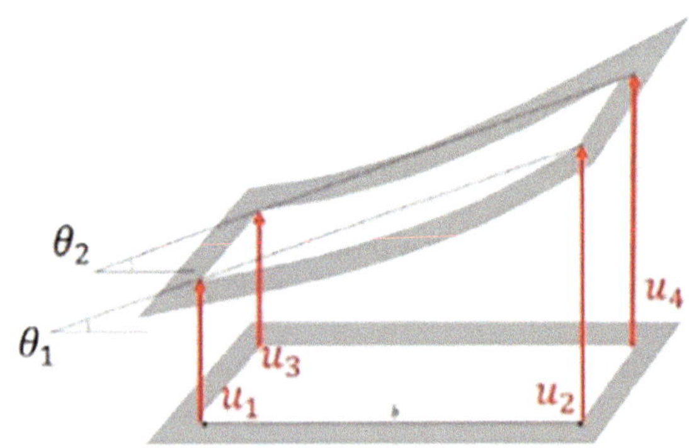

$$\theta = \left| atan\left(\frac{u_{max} - u_{min}}{b}\right) \right|$$

Fig. 10 Main misalignment angles of the collimator frame window averaged to the sides across all 16 panels and establish a single output parameter for minimization during the optimization phase, an average of the 44 respective misalignment angles of the panels was calculated with the formula in Fig. 11. This average value serves as a single metric representing the overall misalignment state of the collimator frame.

In the context of this optimization process, Θ^* denotes the misalignment value calculated relative to the original configuration. Achieving an angle $\Theta \leq \Theta^*$ will be a necessary condition for a successful optimization. In other words, the optimization algorithm aims to minimize the misalignment of the collimator frame to a value less than or equal to the reference misalignment value (Θ^*).

Mesh morphing Building upon the findings from the previous analysis, the RBF Morph tool was employed to perform geometric parameterization. This technique allows for the direct implementation of shape modifiers on the computational domain. The resulting new geometric configurations are achieved through mesh region displacement driven by RBF algorithms. These algorithms smoothly propagate the imposed displacement throughout the surrounding volume.

In this specific case, the RBF Morph tool was used to create three distinct morphing regions, each defining specific displacements within the collimator frame.

- Offset: changes the height of the rib from an initial value of 3 mm to an arbitrary and parametric value.
- Inclination: makes the height of the transverse and longitudinal rods of the rib variable variable, except for the loop in the Fig. 12, whose height is kept constant.
- Thinning: makes the thickness of the internal transverse rods variable from a value of 8 mm towards the center of the collimator at an arbitrary peripheral value, makes the thickness of the rods constant external transversal and lower than the initial one by 8 mm.

Within the simulation software package, specific values were identified to parameterize the RBF Morph tool. These values, defined within the "Parameter Set," correspond to the various configurations (instances) of the morphed mesh.

- h_int: height at the center of the rib.
- h_est: height adjacent to the perimeter of the rib.
- thickness_trasv_int: thickness of the internal transverse rods.
- thickness_trasv_est: thickness of the external transversal rods.

The limits of the optimization parametric domain have been set as:

- h_int: variable between 3 and 5 mm (reference value 3 mm).
- h_est: variable between 1 and 3 mm (reference value 3 mm).
- thickness_trasv_int: variable between 6 and 8 mm (reference value 8 mm).
- thickness_trasv_est: variable between 6 and 8 mm (reference value 8 mm).

Fig. 11 Θ^* formula

$$\theta^{*} = mean(\theta_{11}, \theta_{12}, \ldots, \theta_{44})$$

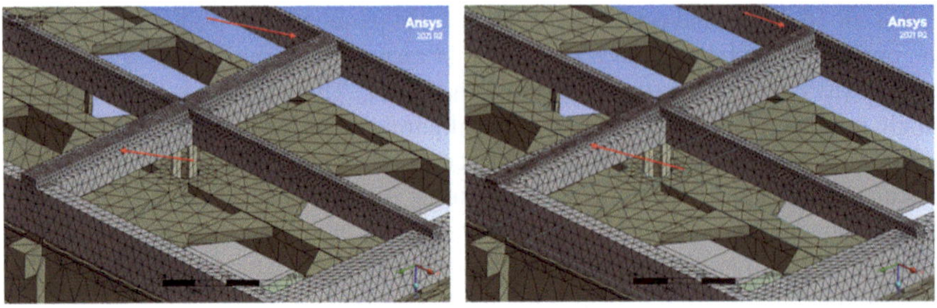

Fig. 12 Rib's morphing

The reference values used for parameterization correspond to the original rib configuration of the simplified model. This optimization problem is formulated as a constrained minimization problem. The objective function minimizes the average misalignment across the panels, while a constraint enforces adherence to the original overall volume of the collimator frame (including the ribbing).

The chosen optimization method, suggested by the OptiSlang tool, is the "Adaptive Response Surface Method." This method iteratively builds a surrogate model of the design space, allowing for efficient exploration and identification of optimal solutions.

While optimizations with large amount of iterations generally lead to more effective designs (with misalignment reduction exceeding 3–5%), computational limitations and time constraints restricted the simulation scope for this study. In other words, increasing the number of iterations would likely result in further improvements, but time and computational resources were limited.

Results Following the optimization process, the resulting design was compared to the original collimator frame based on the misalignment metric. The optimization software was executed for approximately five days across two separate workstations. The results revealed some key findings: to minimize the upward inflection of the collimator frame, increasing the rib height is more effective than increasing its thickness.

In essence, the optimization algorithm identified the optimal design as characterized by obtaining a reduction in misalignment of approximately 5% with the shape obtained in Fig. 13.

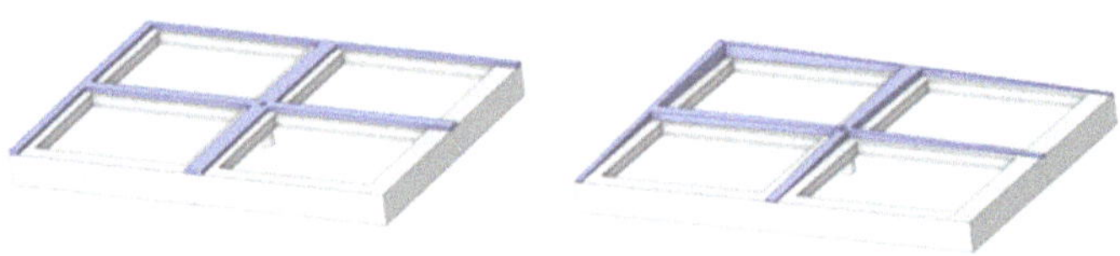

Fig. 13 Original collimator frame (left) versus optimized collimator frame (right)

To maintain the overall volume of the collimator frame, the optimization process strategically increased the rib height in the central region while simultaneously decreasing the height in the peripheral areas. Additionally, the transverse rods were thinned.

The optimized design can be compared to the original design by analyzing the evolution of individual misalignment angles that contribute to the overall misalignment metric. Figure 14 visually demonstrates a significant improvement in the overall distribution of misalignment angles within the collimator frame.

Bipod optimization This section focuses on a second optimization process aimed at the bipod design. The bipods are critical elements that support the module and connect it to the truss structure. The objective of this optimization is to increase the first natural frequency of the system by 15%, targeting a value 120 Hz, which is a mission requirement.

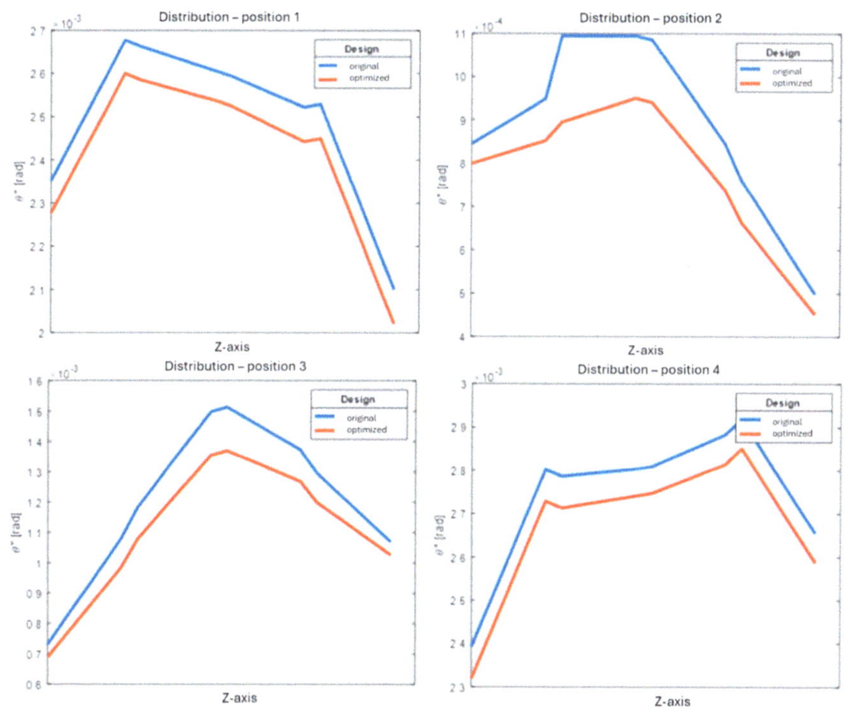

Fig. 14 Comparison of the distributions of the angles of misalignment

The bipods play a crucial role in transmitting inertial and thermal loads from the LAD module to the supporting structure. Consequently, their structural integrity is paramount.

The following steps were undertaken to achieve the desired optimization:

- preliminary design of the module in a complete CAD model meeting the requirements of the mission and export of the model in ANSYS SpaceClaim obtaining a simplified defeatured model but compliant in terms of mass and volume;
- Definition and choice of the bipod's configuration, modal, structural and thermo-elastic simulation.

To achieve the target increase in natural frequency while considering the trade-off between mass and mission cost, three distinct bipod configurations, denoted as A, B, and C, in Fig. 15, were evaluated.

The ANSYS response surface tool was employed to conduct an optimization process on these three configurations. The optimization aimed to minimize the bipod mass while maintaining a natural frequency constraint of greater 138 Hz (which ensures a 15% increase over the original design).

- Configuration A: This design focuses on modulating the thickness of the two bipod walls through parametric variations in height. Additionally, a horizontal insert was added to enhance stiffness.
- Configuration B: Building upon configuration A, this design incorporates a vertical element attached to the horizontal insert to further increase bending stiffness.
- Configuration C: This most complex design utilizes the horizontal insert from configuration B as a base for a reticular structure. The reticular structure is formed by two symmetrical, oblique elements resting on the insert.

To achieve greater stiffness and consequently a higher first natural frequency, the mass of the initial geometry (58.858 g) was systematically increased across different design configurations. Configuration C was ultimately chosen because it offered superior stiffness compared to the baseline design while incurring the least weight penalty.

Having selected the optimal bipod layout, the study proceeded with analyzing the stress state during various satellite operational phases. This involved performing static structural analyses to evaluate the equivalent stresses resulting from the quasi-static loads experienced during the launch phase and temperature differences between various components during the transition from assembly temperature to operational temperature in orbit. The analysis of launch stresses employed acceleration data, in Table 4 from the launch phase in all three directions. The effects were then evaluated in terms of equivalent forces. Conversely, the thermo-elastic stress analysis adopted a simplified approach where individual component temperatures were assigned.

In both cases, the equivalent stress values were extracted according to the von Mises theory.

Table 5 shows the minimum, average and maximum values relating to the equivalent stress for the bipod along the short side of the module, while in figures there

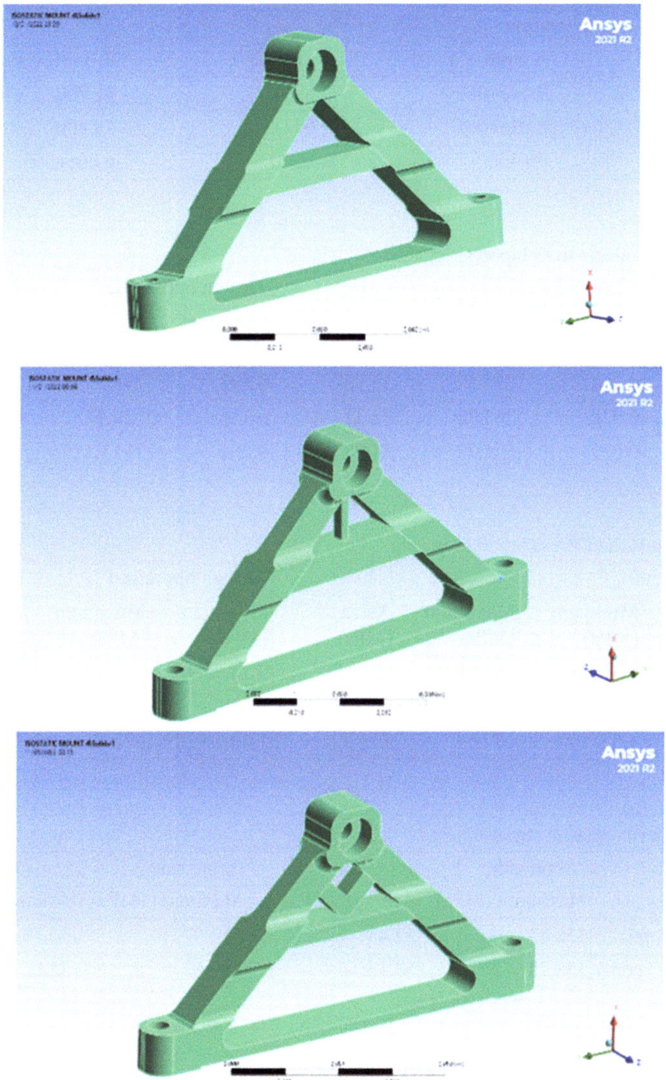

Fig. 15 Configuration A, B and C

is a graphical representation of the instant of time 3 s which is the most burdensome for the bipod.

On the long side, however, the trend of the stress state in the three cases analysed is as follows in Table 6.

To assess the thermo-elastic deformations and verify the structural integrity of the structure, the temperatures corresponding to the most critical thermal case (the cold case) were imported from Table 3 into the finite element model.

Table 4 Launch acceleration

Time (s)	X (mm/s^2)	Y (mm/s^2)	Z (mm/s^2)
1	9.44e+005	2.59e+005	41,200
2	8.03e+005	1.04e+006	79,400
3	79,400	23,500	6.06e+005

Table 5 Equivalent stress bipod short side

Design C conf				Design optimized		
Time (s)	Minimum (MPa)	Maximum (MPa)	Average (MPa)	Minimum (MPa)	Maximum (MPa)	Average (MPa)
1	0.211	272.30	33.08	0.392	251.73	30.463
2	0.928	265.70	42.32	0.346	220.47	41.17
3	0.832	551.70	54.98	0.682	711.92	46.57

Table 6 Equivalent stress bipod long side

Design C conf				Design optimized		
Time (s)	Minimum (MPa)	Maximum (MPa)	Average (MPa)	Minimum (MPa)	Maximum (MPa)	Average (MPa)
1	0.175	251.28	29.70	9.74e−002	259.37	26.82
2	0.873	524.57	59.50	0.618	536.14	55.35
3	1.52	677.57	72.89	1.43	677.11	61.44

Table 7 Thermo-elastic stress

Case	Short side		Long side	
	Maximum (MPa)	Average (MPa)	Maximum (MPa)	Average (MPa)
Cold case (opti)	458.85	52.43	312.07	30.18
Cold case (C conf)	588.75	58.13	299.67	32.6

The reference temperature for component assembly was set at 20 °C.

Table 7 summarizes the average and maximum equivalent stress values obtained through static structural analysis. These values were calculated considering the thermal loads applied to the module components for both the original and optimized designs.

Results The bipod design was optimized to meet the frequency requirements of the first vibration mode. This work evaluated different designs with the goal of achieving the desired frequency increase while minimizing weight gain.

The target frequency for the first significant vibration mode 120 Hz, with a 15% margin. This study successfully increased the natural frequency from the initial value of 122.69 to 138.05 Hz (a + 12.5% increase), but with a corresponding mass increase of 10 g per bipod (a + 17.2% increase).

Following the optimization, the stress state of the new bipod layout was evaluated. As expected with a design optimized for frequency, some "hot spots" were identified, leading to higher maximum and average stress values compared to the original design. However, the resulting stress levels remain well below the Aluminium material's yield limits, ensuring the structure remains within the elastic range.

4 CUSP—CubeSat Solar Polarimeter

Our understanding of the Universe is primarily derived from the analysis of electromagnetic radiation received on Earth from astronomical objects. Solar flares (SFs) are eruptive events on the Sun that release intense electromagnetic radiation and energetic particles. These events can significantly disrupt human activities on the ground and in space. Solar activity, including SFs, can degrade radio communication performance, cause radio blackouts, and interfere with GPS and satellite navigation systems. Additionally, high-energy particles (protons and electrons) released during SFs can deposit their energy in satellite electronics, leading to malfunctions or even complete satellite loss. The occurrence of SFs is also correlated with Coronal Mass Emissions (CME) and Solar Energetic Particles (SEP) impacting ground infrastructure [16].

The CUSP mission, shown in Fig. 16, aims to contribute to existing and future space weather monitoring networks, including the future SPace weather InfraStructure (ASPIS) mission by the Italian Space Agency (ASI) [17].

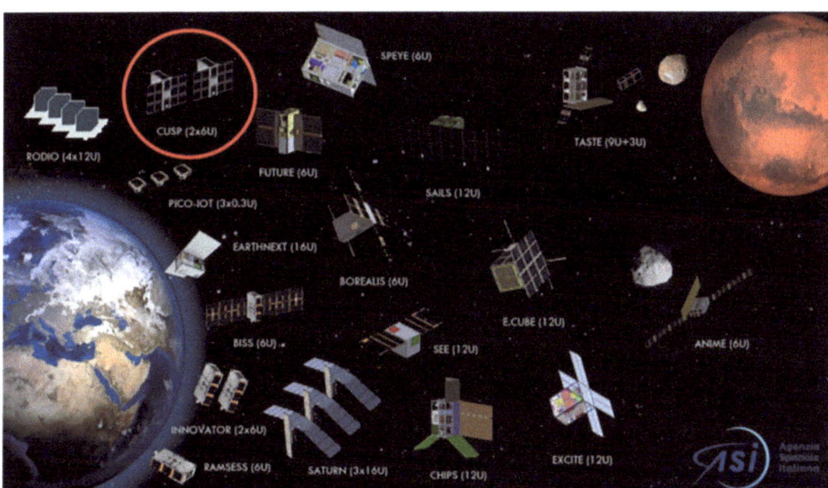

Fig. 16 CubeSat technologies and missions

CUSP is currently undergoing a Phase A study funded by ASI within the Alcor program, which focuses on the development of CubeSat technologies and missions. CUSP was one of 20 proposals selected from a pool of 49 submitted for the CubeSat mission call. This call for proposals attracted participation from 22 research institutions and universities, alongside 78 companies, primarily Small and medium-sized enterprises (SMEs).

The Italian National Institute for Astrophysics—Institute of Astrophysical Space Physics (INAF-IAPS) is the prime contractor for CUSP and is responsible for developing the scientific payload.

To maximize the scientific and application value of the mission, CUSP requires continuous solar observation. This requirement presents a significant challenge for a CubeSat mission due to the launch typically being secondary to a primary payload that dictates orbital parameters. To address this challenge, the project has opted for a constellation of two satellites equally spaced within the same orbital plane.

This constellation configuration ensures that at least one satellite is constantly exposed to the Sun, allowing it to point its payload directly towards the Sun. The target orbit for the constellation is a Sun-synchronous orbit with a Local Time of Ascending Node (LTAN) ranging from a Mid-Morning scenario (LTAN = 9:30) to a Noon-Midnight scenario (LTAN = 12:00), with an altitude between 500 and 600 km.

4.1 Mission Concept and Platform

The project foresee a constellation consisting of one orbital plane and two CubeSats at 180° of phase difference along the orbit shown in Fig. 17. This configuration allows to have always at least 1 satellite of the constellation in daylight for any β angle (orientation of the orbital plane with respect to the Sun) for observing the Sun.

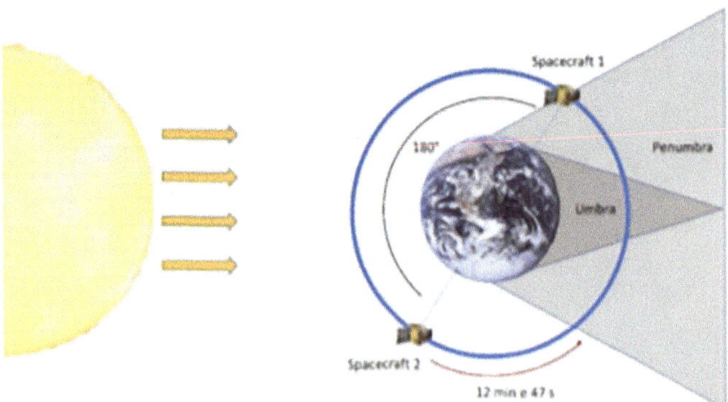

Fig. 17 Constellation of two CubeSats at 180° phase difference along the orbit

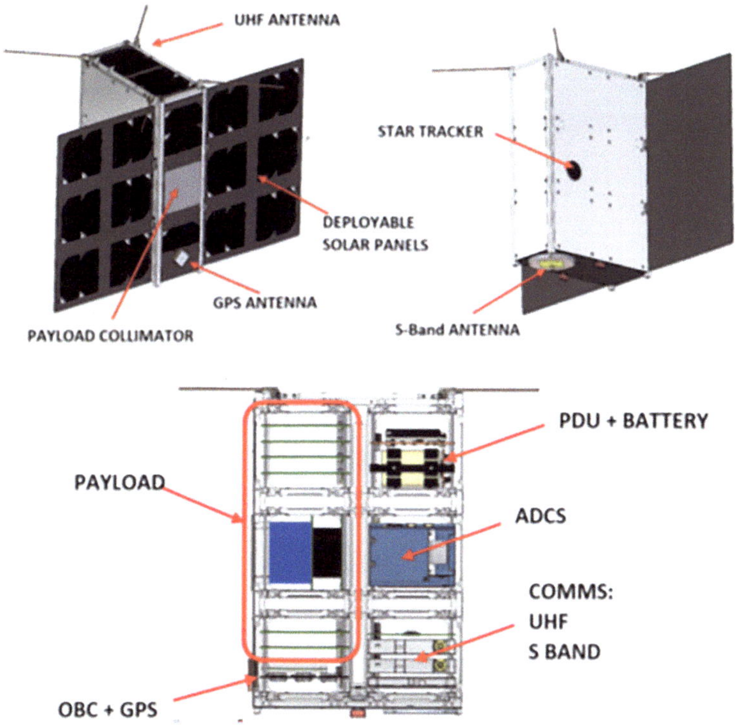

Fig. 18 CUSP platform and subsystem

The time scale of the project, until launch, and then the operational phase of the mission, are compatible with the observation of the Sun during the next cycle maximum.

The platform, in Fig. 18, of the CUSP project is a standard 6U CubeSat architecture developed by IMT s.r.l., that has a consolidated experience in the realization of different CubeSats, also in the institutional framework.

4.2 *Mechanical Design*

The CUSP payload, Fig. 19, consists of two main trays: a collimator tray and a sensor tray. These trays house the components responsible for collecting and measuring solar radiation. During the mission's Phase A, where the design is being finalized, mechanical and electrical features like connectors, threads, and screws are not yet included.

The collimator tray houses eight tungsten alloy collimators, four each for the scatterers and absorbers. These collimators are positioned to correspond with individual

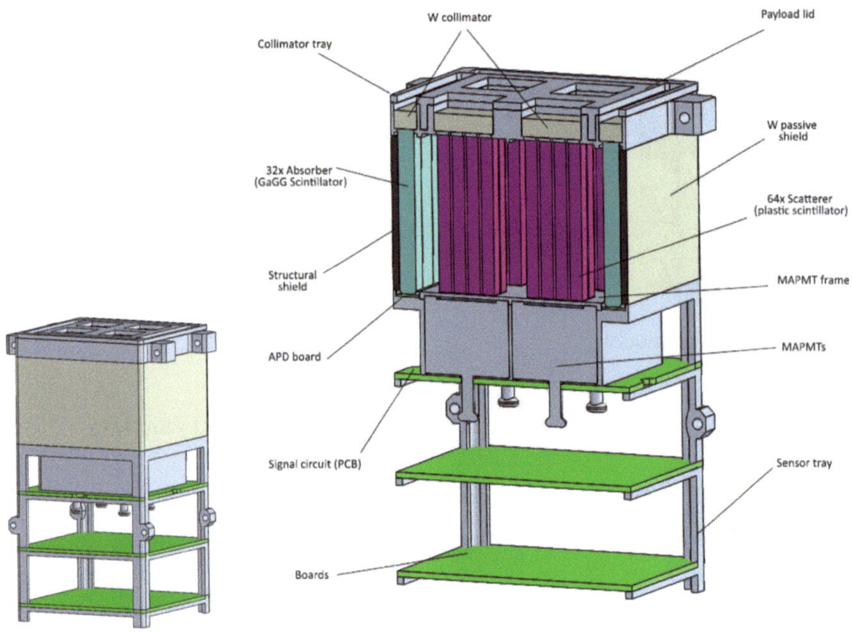

Fig. 19 Compact (on the left) and exploded view of the hard X-ray polarimeter of the CUSP payload

MAPMT detectors and APD boards. The tray is topped with an aluminum (anticorodal) lid designed for space applications. This lid is clamped onto a thin ($\sim$1 μm) passive shield.

Figure 20 shows the CUSP lid and illustrates both absorber and scatterer collimators. Each scatterer collimator is $18 \times 18 \times 5\,\text{mm}^3$ in size, while the absorber collimators are larger at $58.4 \times 7 \times 5\,\text{mm}^3$. The absorber collimators fit into recesses within the aluminum alloy collimator frame. The collimator frame also includes external ties that allow it to be mounted onto the CubeSat truss.

The sensor tray houses the APD boards and the MAPMT frame in its upper part. Each of these boards has a mounted scatterer and absorber scintillator. The scatterer, made of plastic scintillator (EJ404), has dimensions of $3.7 \times 3.7 \times 45\,\text{mm}^3$. The absorber, a GaGG scintillator, is slightly larger at $5 \times 7 \times 45\,\text{mm}^3$. Figure 21 shows an exploded view of the sensor tray and depicts both absorber and scatterer scintillators.

The payload also includes a high voltage (HV) system, a micro power supply, and other electronic components housed on boards. One of these boards is specifically linked to the MAPMT.

Overall, the CUSP payload utilizes a tungsten collimator to restrict the field of view to approximately ± 21 degrees around the solar direction. The payload then employs both plastic and inorganic scintillators for the detection of solar radiation. These scintillators are read out by 4 MAPMTs (providing a total of 64 channels) and 32 APDs. The selection of high heritage readout sensors (MAPMTs and APDs)

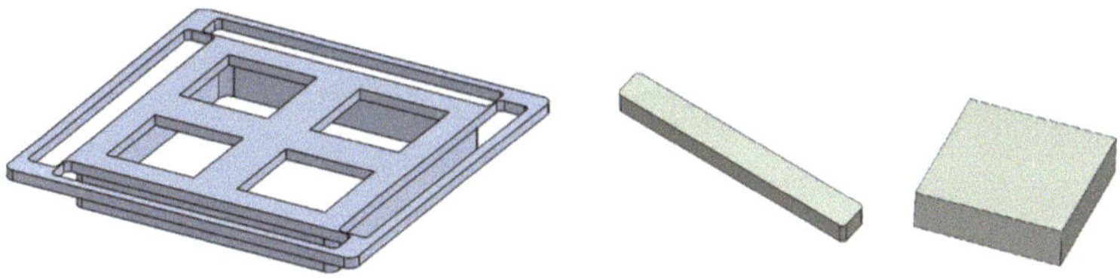

Fig. 20 CUSP lid (left), absorber (center) and scatterer (right) collimator

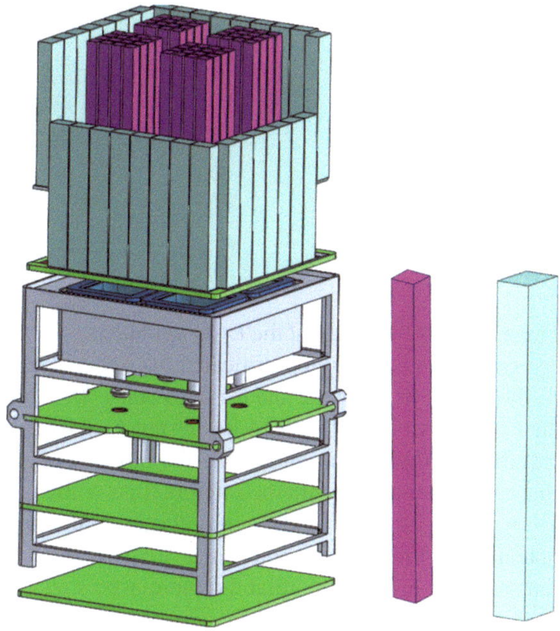

Fig. 21 Exploded view of the sensor tray (left) and plastic/GaGG scintillators (right)

minimizes development risks and ensures reliable operation within the project's short implementation timeframe.

5 Satellite Interface

The interface, in Fig. 22, between the CUSP payload and the host 6U CubeSat is achieved by utilizing eight interface tiles. Four tiles are strategically mounted on the sensor tray, while the remaining four are positioned on the collimator frame. This configuration ensures the precise relative positioning of the payload's instrumentation with respect to the satellite structure, facilitating a proper and secure integration of CUSP within the CubeSat.

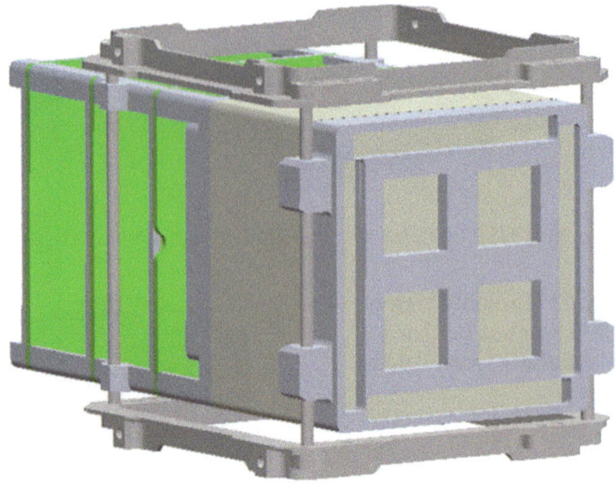

Fig. 22 CUSP in the structure's basket

6 Conclusions

The multi-physics study of the LAD module is ongoing, accompanied by concurrent design studies focusing on critical subsystems. These subsystems, particularly the collimator, detector plate, and truss interface (I/F), present unique challenges due to their complexity and the need for design trade-offs and adaptations. Critical areas requiring further definition include the bipod configuration (position, quantity, and shape) for connecting with the satellite truss while maintaining the MPO alignment.

Dynamic vibration and thermo-elastic analyses will be conducted to identify potential optimization opportunities for structural frames and connectors. Material selection prioritizes initial thermo-elastic stress reduction and high strength for long-term durability. Optimizing these parameters is expected to significantly reduce stresses and dynamic displacements.

The collimator frame optimization exemplifies the multi-disciplinary design approach. A thermo-elastic structural analysis was initially performed for the cold thermal case. This analysis revealed the stresses and deformations experienced by the initial geometry, paving the way for potential structural optimization.

The model was subsequently modified to include additional processing parameters as optimization variables. The collimator misalignment angle was designated as the merit parameter for a geometric optimization leveraging ANSYS tools. The candidate design achieved a successful misalignment reduction of approximately 5% compared to the initial configuration. Future efforts will focus on developing solutions for complete decoupling of the collimator frame, enabling direct misalignment angle calculations on the collimators themselves. This approach was initially disregarded to minimize computational burden.

A similar optimization approach was applied to the bipod design. A minimum required natural vibration frequency was imposed as a constraint, incorporating a 15% safety factor, resulting in a threshold value 138 Hz.

The initial design exhibited a natural frequency of 122 Hz, falling short of the imposed constraint. Consequently, the support topology was redesigned by strategically adding material to enhance stiffness in areas exhibiting high flexibility. The optimized topology boasts a 17.2% mass increase compared to the original design but fulfills the critical natural frequency requirement, achieving a value of 138.05 Hz. Following the definition of the candidate topology, successful verification for all required load conditions was performed. As anticipated, the additional material in the optimized design led to reduced stresses under thermal and launch load conditions, ensuring all values remain within the material yield point.

As previously mentioned, an high fidelity CAE study is crucial for CUSP due to the complex interplay of various physical phenomena within the spacecraft.

The mechanical design of CUSP is progressing steadily, with in-depth studies focusing on critical subsystems like the detector and collimator (including position, quantity, and shape) and their integration with the satellite truss. Defining the optimal configuration for these elements remains a challenge due to the interplay of multiple requirements.

Future endeavors include static and dynamic numerical simulations along with thermo-elastic analyses (both steady-state and transient) to identify potential optimization opportunities for structural frames and connectors. Material selection has prioritized preliminary mitigation of thermo-mechanical-elastic stresses while ensuring high structural strength and long-term durability.

A key future effort for CUSP involves conducting a detailed thermal simulation using specialized software. This simulation will encompass a detailed model of a single payload module alongside the overall CubeSat structure.

References

1. Neelamkavil, F.: Computer Simulation and Modelling. Wiley (1987)
2. Zhang, S., Santangelo, A., Feroci, M., Xu, Y., et al.: The Enhanced X-ray Timing and Polarimetry Mission, Inc. (2019)
3. Feroci, M., Ahangarianabhari, M., Ambrosi, G., Argan, A., et al.: The Large Area Detector Onboard the eXTP Mission (2018)
4. Hernanz, M., Brandt, S., Feroci, M., Orleansky, P., Santangelo, A., et al.: The Wide Field Monitor Onboard the eXTP Mission (2018)
5. Weisskopf, M., et al.: Proc. SPIE **9905**, 990517 (2016)
6. Mendes, V.: OAC INAF/IAPS Italy, S. Thermal Analysis of the eXTP-LAD Instrument (2022)
7. Biancolini, M.E.: Fast Radial Basis Functions for Engineering Applications. Springer (2017)
8. Pascoletti, G., Bignardi, M., Conti, C., Zanetti, P.: Mandible Morphing Through Principal Components Analysis Applications. Springer (2020)
9. Cali, M., Oliveri, S.M.: Application of an effective SIMP method with filtering for topology optimization of motorcycle tubular frame. Int. Rev. Mech. Eng. (2017)
10. Di Domenico, N., Groth, C., Wade, A., Berg, T., Biancolini, M.E.: Fluid structure interaction analysis. Integrity (2018)
11. Cali, M., Speranza, D., Cella, U., Biancolini, M.E.: Flying Shape Sails Analysis by Radial Basis Function Mesh Morphing. Springer (2020)
12. Zienkiewicz, O.C., Taylor, R.L., Zhu, J.Z.: The Finite Element Method: Its Basis and Fundamentals. Butterworth-Heinemann (2005)

13. Dassault Systemes. ANSYS OptiSLang. www.ansys.com/products/connect/ansys-OptiSlang
14. Dynardo, D.S.: Solidworks 2022. https://www.solidworks.com/it
15. Airbus. Systema Thermica. https://www.airbus.com/en/products-services/space/customer-services/systema
16. Papaioannou, A., Sandberg, I., Anastasiadis, A., Kouloumvakos, et al.: Solar flares, coronal mass ejections and solar energetic particle event characteristics. Space Weath. Space Climate (2016)
17. Plainaki, C., Negri, B., Castronuovo, M., Antonelli, A.: Towards an Italian space weather infrastructure: the ASPIS Project. In: EGU General Assembly Conference Abstracts (2018)
18. Hajdas, W., Suarez-Garcia, E.: Polarimetry at high energies. In: Observing Photons in Space (2018)

MIX
Papier aus verantwortungsvollen Quellen
Paper from responsible sources
FSC® C105338

If you have any concerns about our products,
you can contact us on
ProductSafety@springernature.com

In case Publisher is established outside the EU,
the EU authorized representative is:
Springer Nature Customer Service Center GmbH
Europaplatz 3, 69115 Heidelberg, Germany

Printed by Libri Plureos GmbH
in Hamburg, Germany